T0262360

Encyclopedia of Recrystallization: Metallic Minerals

Volume I

Encyclopedia of Recrystallization: Metallic Minerals
Volume I

Edited by **Sylvia Dickey**

New York

Published by NY Research Press,
23 West, 55th Street, Suite 816,
New York, NY 10019, USA
www.nyresearchpress.com

Encyclopedia of Recrystallization: Metallic Minerals
Volume I
Edited by Sylvia Dickey

International Standard Book Number: 978-1-63238-163-7 (Hardback)

Printed in the United States of America.

Contents

Preface   VII

Recrystallization of Metallic Materials   1

Chapter 1   **Recrystallization of**
             Dispersion-Strengthened Copper Alloys   3
             Su-Hyeon Kim and Dong Nyung Lee

Chapter 2   **Development of Texture and Substructure Inhomogeneity**
             by Recrystallization of Rolled Zr-Based Alloys   23
             Yuriy Perlovich and Margarita Isaenkova

Chapter 3   **Crystal Growth:**
             Substructure and Recrystallization   43
             Vadim Glebovsky

Chapter 4   **Application of Orientation Mapping in TEM and**
             SEM for Study of Microstructural Evolution
             During Annealing – Example: Aluminum Alloy
             with Bimodal Particle Distribution   71
             K. Sztwiertnia, M. Bieda and
             A. Kornewa

Chapter 5   **Physical Metallurgy and Drawability of**
             Extra Deep Drawing and Interstitial Free Steels   87
             Kumkum Banerjee

Chapter 6   **The Failure Mechanism of Recrystallization –**
             Assisted Cracking of Solder Interconnections   129
             Toni T. Mattila and Jorma K. Kivilahti

Chapter 7   **Phase Transformations and Recrystallization Processes**
             During Synthesis, Processing and Service of TiAl Alloys   157
             Fritz Appel

Chapter 8   **Mathematical Modeling of Single Peak Dynamic
             Recrystallization Flow Stress Curves in Metallic Alloys**   **199**
             R. Ebrahimi and E. Shafiei

Chapter 9   **The Effect of Strain Path on the Microstructure and
             Mechanical Properties in Cu Processed by COT Method**   **217**
             Kinga Rodak

             Permissions

             List of Contributors

Preface

This book is a compilation of researches conducted by scientists on issues related to recrystallization. Various disciplines of science, like geology, metallurgy etc., in which recrystallization plays an essential role have been covered. The phenomenon of recrystallization has seen constant growth in the last few decades due to the researches being conducted in various branches of science. This book provides a detailed analysis of recrystallization of metallic minerals and discusses proper techniques and methods of characterization. Researchers, students and scientists would benefit from this all encompassing book.

This book is a comprehensive compilation of works of different researchers from varied parts of the world. It includes valuable experiences of the researchers with the sole objective of providing the readers (learners) with a proper knowledge of the concerned field. This book will be beneficial in evoking inspiration and enhancing the knowledge of the interested readers.

In the end, I would like to extend my heartiest thanks to the authors who worked with great determination on their chapters. I also appreciate the publisher's support in the course of the book. I would also like to deeply acknowledge my family who stood by me as a source of inspiration during the project.

Editor

Recrystallization of Metallic Materials

Recrystallization of Dispersion-Strengthened Copper Alloys

Su-Hyeon Kim[1] and Dong Nyung Lee[2]
[1]Korea Institute of Materials Science,
[2]Seoul National University[2]
Republic of Korea

1. Introduction

1.1 Dispersion-strengthened copper alloys

Pure copper exhibits high electrical and thermal conductivities, but it has low strength at room temperature as well as at elevated temperatures. Dispersion-strengthened (DS) copper alloy exhibits a high strength without sacrificing its inherent high conductivities, and maintains excellent thermal and mechanical stability at elevated temperatures by retaining its microstructures (Nadkarni, 1984). These unique characteristics are mainly attributed to the presence of uniformly dispersed thermally stable particles, which are typically oxides. Unlike precipitation-hardened copper alloys, which lose their strength by heating above the initial aging temperatures, the non-metallic oxide particles in oxide DS copper alloys, such as alumina, silica, and beryllia, neither coarsen nor go into solution, effectively preventing recrystallization and consequent softening of the alloys. Alumina DS copper alloys are not recrystallized even after exposure to temperatures approaching the melting point of copper (Preston & Grant, 1961). This is due to the pinning effect of the nano-sized alumina particles on the movement of the boundaries and dislocations. A unique combination of high strengths and high conductivities at elevated temperatures makes alumina DS copper alloys good candidates for high temperature electric materials (e.g., electrodes, lead wires, and connectors) (Nadkarni, 1984) as well as potential components in nuclear energy applications (Sumino et al., 2009).

Alumina DS copper alloys can be recrystallized when boron is added (Kim & Lee, 2001, 2002). Boron is often intentionally added as an oxygen scavenger during fabrication of the alloys (Gallagher et al., 1992). Long term annealing of boron-added alumina DS copper alloys results in an unexpected transformation from fine γ-Al_2O_3 to coarse $9Al_2O_3$-$2B_2O_3$ with a concurrent recrystallization of the matrix to form a large and elongated grain structure (Kim & Lee, 2002). Whereas Ni-based DS alloys are used in a coarse-grained condition to increase high-temperature creep resistance (Gessinger, 1976; Stephens & Nix, 1985), key applications of alumina DS copper alloys require them to be in a fully work-hardened state. Consequently, a large decrease in room temperature strength due to recrystallization is not desirable. Therefore, an understanding of the recrystallization behaviour of DS copper alloys is important from both practical and theoretical perspectives.

1.2 Recrystallization of particle-containing alloys

The presence of dispersed particles critically affects the plastic deformation and recrystallization behaviour of the matrix. The presence of particles accelerates or retards recrystallization of the matrix, depending on the interparticle spacing, size, mechanical properties, and thermal stability of the particles (Humphreys & Hatherly, 1995). Closely spaced fine particles exert a pinning effect on the movement of boundaries (Zener drag) resulting in retardation or even complete suppression of recrystallization. However, alloys with widely spaced particles larger than ~1 μm show accelerated recrystallization. Non-deformable large particles can introduce deformation zones around the particles during deformation, providing favourable nucleation sites for recrystallization (particle stimulated nucleation, PSN). Under certain conditions, particle-containing alloys transform from a deformed structure to a recrystallized grain structure in the absence of conventional discontinuous recrystallization accompanying a long-range motion of the boundaries. During low-temperature annealing, small particles give rise to boundary pinning, and subsequent coarsening of the particles at high temperatures may allow the subgrains to grow, forming recrystallized grain structures. This phenomenon is sometimes known as continuous recrystallization.

1.3 Purpose of the study

While several studies exist on the fabrication methods, mechanical properties, and deformation behaviour of alumina DS copper alloys, there is a lack of understanding of their recrystallization behaviour. This study examines the recrystallization behaviour of boron-added alumina DS copper strips rolled under different conditions. Particular attention is given to several anomalous phenomena, such as unique recrystallized grain structures and textures, as well as the dependency of recrystallization characteristics on prior rolling conditions. The results of several microscopy studies to elucidate microstructural evolution during rolling and annealing are presented, and the effects of dispersed particles on recrystallization are examined.

2. Research methods

2.1 Materials

The material used in this study was commercially available alumina DS copper alloy strips, Glidcop Al25, produced by SCM Metal Products. This material contains 0.25wt% Al in the form of Al_2O_3 particles as well as 0.02wt% B used for oxygen scavenging. The thickness of the as-received strips was 840 μm. The chemical composition of the as-received strips was measured by inductively coupled plasma (ICP) analysis and given in Table 1.

Al	B	P	Fe	S	As	Mn	Cu
0.275	0.023	0.0001	0.0001	0.0001	0.0034	0.0001	Balance

Table 1. Chemical composition of the as-received strips measured by ICP (wt%)

2.2 Rolling and annealing

The as-received strips were rolled under lubrication using a two-high rolling mill whose roll diameter was 126 mm to make two different specimens, as listed in Table 2. The cold-rolled strips were further rolled to reduce their thickness by 25% with one pass at room temperature. The thickness of the hot-rolled strips was reduced by 27% with one pass after heating the strips at 813 K for 10 minutes. Isothermal annealing of the as-received and rolled strips was carried out in a salt bath. After the heating, the strips were quenched in water.

Specimen	Number of passes	Total reduction	Rolling temperature	Lubrication
Cold-rolled strip	1	25%	Ambient	Yes
Hot-rolled strip	1	27%	813 K	Yes

Table 2. Rolling conditions of the as-received strips

2.3 Microstructure and texture analysis

The microstructures of the strips were investigated by optical microscopy and transmission electron microscopy (TEM) in the transverse direction (TD) and the normal direction (ND). The specimens were cut from the strips, mechanically polished, and chemically etched in $FeCl_3$ solution prior to optical microscopy. For the TEM study, the specimens were electrically polished in a nitric acid solution to make a thin foil using a twin-jet electro-polisher, while the dispersed alumina particles were extracted from the material using a carbon replica method.

The macroscopic textures of the strips were determined by measuring (111), (200), and (220) pole figures with an X-ray diffraction goniometer in the back reflection mode with Co $K\alpha$ radiation. The specimens were mechanically polished parallel to the rolling plane and chemically etched in a nitric acid solution. Three-dimensional orientation distribution functions (ODFs), complete pole figures, and orientation densities were calculated from the measured pole figures using the WIMV program (Matthies et al., 1987). The orientations of individual crystallites were calculated from the Kikuchi patterns obtained by TEM (Young, et al. 1973) in the TD section of the specimens. Misorientations between adjacent crystallites were calculated using 24 symmetry operations (Randle, 1993).

2.4 Analysis of the mechanical properties

Tensile tests of specimens with a gauge length of 30 mm along the rolling direction (RD) were carried out at room temperature at a crosshead speed of 1 mm/min. The micro-Vickers hardness of the specimens was measured under a load of 25 g for 10 s.

3. Results

3.1 Characterization of the as-received strips

Figure 1 shows the microstructures of the as-received strips observed under an optical microscope. The material exhibited a highly deformed microstructure consisting of fine band-like substructures aligned nearly parallel to the RD. Figure 2 shows the longitudinal section TEM microstructure observed in the surface and centre regions of the as-received strips. The average band thicknesses of the surface and the centre regions were 0.127 and 0.129 μm, respectively. Additional band boundary characteristics measured on the centre region are given in Table 3. The grain structure of the as-received strips was characterized by a fine band-like grain structure with a high-angle boundary character.

The mechanical properties of the as-received strips are given in Table 4. The high strengths and hardness indicate that the strips were heavily deformed.

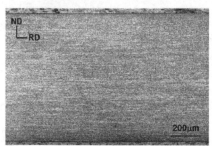 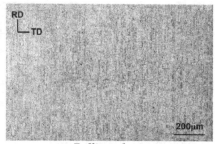

| Longitudinal section | Rolling plane view |

Fig. 1. Optical micrographs of the as-received strips

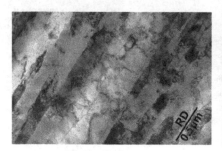 

| Surface region | Centre region |

Fig. 2. Longitudinal section TEM micrographs of the as-received strips

Specimen	Average band thickness (μm)	Average boundary misorientation (deg)	High angle boundary fraction (misorientation ≥ 15 deg)
As-Received	0.129	30.6	0.52

Table 3. Band boundary characteristics of the as-received strips

Specimen	Tensile strength (MPa)	Yield strength (MPa)	Elongation (%)	Hardness (Hv)
As-Received	553	515	14	169

Table 4. Mechanical properties of the as-received strips

Figure 3 shows the texture evolution of the as-received strips. The texture was characterized by the β-fibre, running from the copper orientation {112}<111> over the S orientation {123}<634> to the brass orientation {011}<211> in the Euler orientation space. The well-developed β-fibre texture indicated that the received strip was in a heavily rolled state, which is consistent with the microstructure evolution shown in Figures 1 and 2. Figure 4 shows the orientation densities along the β-fibre of the surface and the centre regions. The orientation densities of the brass and the S components were higher than the copper component, which is unlike plane-strain rolled pure copper sheets where the copper component is dominant (Hirsch & Lücke, 1988). Also noteworthy is the fact that the density of the brass component was lower than that of the S component in the surface region, while the brass and the S components were almost equally dominant in the centre region.

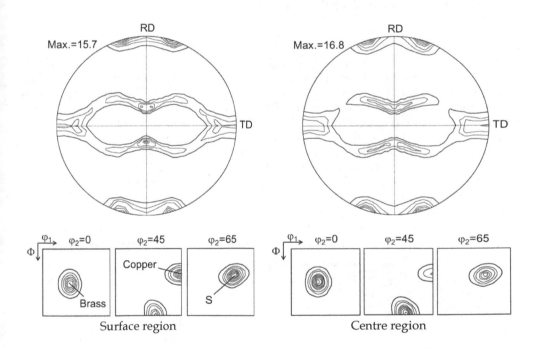

Fig. 3. (111) pole figures and ODFs of the surface and the centre layers of the as-received strips (Kim & Lee, 2002)

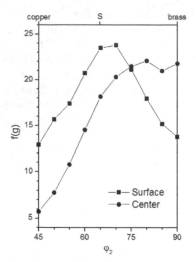

Fig. 4. Orientation densities along the β-fibre of the surface and the centre regions of as-received strips

Figure 5 shows optical microstructures of the as-received strips annealed at 1123 K for 1 hr. Recrystallization occurred in the centre region while no recrystallization took place in the surface region. The plate-like morphology of the recrystallized grains and the ragged shape of the grain boundaries are similar to other extruded or rolled dispersion-strengthened alloys after recrystallization (Klug et al., 1996; Chou, 1997). The TEM microstructures of a recrystallized grain (Figure 6) show dispersed particles aligned parallel to the rolling direction in the recrystallized regions. The micro-Vickers hardness values of the centre and the surface regions were 136 and 168, respectively. This result indirectly indicates that the centre region was recrystallized but the surface region was not. Figure 7 shows the TEM plane view observation of a recrystallized grain in the centre region. A large plate-like recrystallized grain was identified. The textures of the annealed strips are visible in Figure 8. The surface region retained the β-fibre texture, which was similar to the texture in the rolled state. The centre region exhibited a strong texture component, which could be approximated by {112}<312>. The texture of the centre region originated from the recrystallized grains.

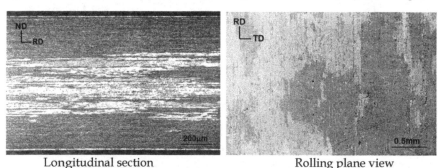

Longitudinal section   Rolling plane view

Fig. 5. Optical micrographs of the as-received strips annealed at 1123 K for 1 hr

Surface region   Centre region

Fig. 6. Longitudinal section TEM micrographs of the as-received strips annealed at 1123 K for 1 hr

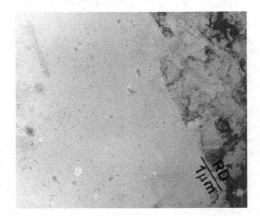

Fig. 7. Rolling plane view TEM micrographs of the centre region of the as-received strips annealed at 1123 K for 1 hr

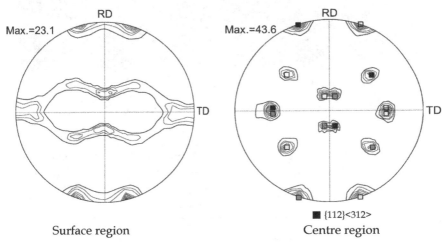

■ {112}<312>

Surface region   Centre region

Fig. 8. (111) pole figures of the as-received strips annealed at 1123 K for 1 hr (Kim & Lee, 2002)

3.2 Properties of the rolled strips

Figures 9 and 10 show optical and TEM microstructures of the cold-rolled and hot-rolled strips. Band-like structures aligned parallel to the RD were observed that were similar to those of the as-received strips. No dynamically recrystallized grains were found in the hot-rolled strips. Table 5 details the band structure characteristics of the cold-rolled and hot-rolled strips measured in the centre regions of each strip. By comparing the band structure characteristic of the as-received strips given in Table 3, the thickness of the band was decreased by cold rolling and increased by hot rolling. Cold rolling also increased the high-angle boundary fraction.

Table 6 shows the mechanical properties of the rolled strips. The cold-rolled strip showed higher strengths and hardness than the hot-rolled strip. By comparing the properties to those of the as-received strips, it can be seen that both cold rolling and hot rolling increased the strengths and hardness of the strips while decreasing their elongation.

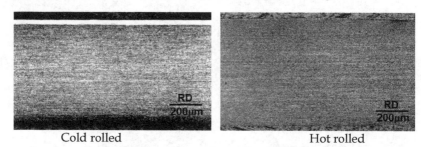

Cold rolled   Hot rolled

Fig. 9. Longitudinal section optical micrographs of the rolled strips (Kim & Lee, 2002)

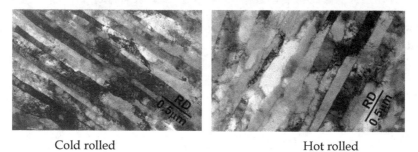

Cold rolled   Hot rolled

Fig. 10. Longitudinal section TEM micrographs of the centre region of the rolled strips (Kim & Lee, 2002)

Specimen	Average band thickness (μm)	Average boundary misorientation (deg)	High angle boundary fraction (misorientation ≥ 15 deg)
Cold rolled	0.116	27.9	0.60
Hot rolled	0.141	24.5	0.51

Table 5. Band structure characteristics of the cold-rolled and hot-rolled strips

Specimen	Tensile strength (MPa)	Yield strength (MPa)	Elongation (%)	Hardness (Hv)
Cold rolled	605	579	5	184
Hot rolled	580	553	5.5	179

Table 6. Mechanical properties and hardness of the cold-rolled and hot-rolled strips

The textures of the rolled strips were similar to those of the as-received strips. Figure 11 shows the orientation densities along the β-fibre of the surface and the centre regions of the rolled strips. The textures of the rolled strips were characterized by the strong β-fibre.

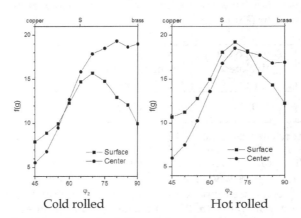

Cold rolled   Hot rolled

Fig. 11. Orientation densities along the β-fibre of the surface and centre regions of the rolled strips

Figure 12 shows the optical microstructures of the rolled strips annealed at 1123 K for 1 hr. TEM micrographs of the centre region are given in Figure 13. Similar to the as-received strips, the cold-rolled strips exhibited recrystallization in the centre region while the hot-rolled strips did not show recrystallization since no recrystallized grains were observed throughout the examined area. However, substantial band growth appeared on the hot-rolled strip. Table 7 gives the band structure characteristics of the hot-rolled and annealed strips. By comparing the results in Tables 5 and 7, it appears that annealing increased the band thickness and high-angle boundary fraction of the hot-rolled strip.

Cold rolled and annealed   Hot rolled and annealed

Fig. 12. Longitudinal section optical micrographs of the rolled strips annealed at 1123 K for 1 hr (Kim & Lee, 2002)

Cold rolled and annealed   Hot rolled and annealed

Fig. 13. Longitudinal section TEM micrographs of the centre region of the rolled strips annealed at 1123 K for 1 hr (Kim & Lee, 2002)

Specimen	Average band thickness (μm)	Average boundary misorientation (deg)	High angle boundary fraction (misorientation ≥ 15 deg)
Hot rolled and annealed	0.270	36.0	0.81

Table 7. Band structure characteristics of the hot-rolled strip annealed at 1123 K for 1 hr

Figure 14 shows (111) pole figures of the cold-rolled strip annealed at 1123 K for 1 hr. The texture of the surface region was characterized by the β-fibre, and the recrystallization texture in the centre region was indexed by {112}<312>. The texture of the hot-rolled strip after annealing is shown in Figure 15. Both the surface and the centre regions retained most of the β-fibre rolling texture.

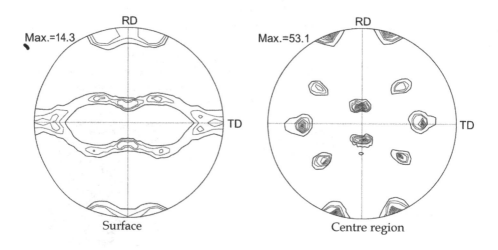

Surface   Centre region

Fig. 14. (111) pole figures of the cold-rolled strip annealed at 1123 K for 1 hr (Kim & Lee, 2002)

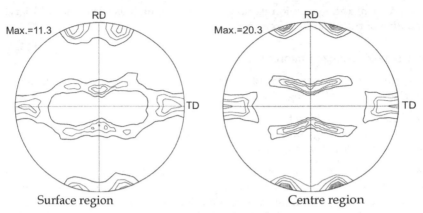

Fig. 15. (111) pole figures of the hot-rolled strip annealed at 1123 K for 1 hr (Kim & Lee, 2002)

4. Discussion

Earlier studies (Preston & Grant, 1961; Nadkarni, 1984) have shown that alumina DS copper alloys resist recrystallization up to their melting points due to the presence of thermally stable alumina particles. The present study showed that alumina DS copper alloys recrystallized after moderate-temperature annealing when boron was added. This is attributed to a reduction in the particle-pinning effect caused by the transformation of particles from fine alumina to coarse aluminium boron oxide. Additionally, large particles already present in the deformed state can introduce deformation zones that act as nucleation sites for recrystallization. Alumina DS copper alloys are fabricated by internal oxidation of Cu–Al alloy powders, consolidations of the powders into fully dense shapes, and further cold rolling to final shapes. The internal oxidation involves the mixing and heating of the alloy powders with oxidants like Cu_2O. Frequently, residual oxygen, or unconverted Cu_2O, may react with hydrogen introduced during alloy processing. This produces a large internal pressure of water vapour and results in blister formations. The material used in this study, Glidcop, is made oxygen-free by intentionally adding boron as an oxygen scavenger. Figure 16 shows a coarse particle observed in the as-received strips, which was identified as $9Al_2O_3$–$2B_2O_3$ by indexing its diffraction patterns. Phase transformation of the particles is expected during the fabrication of a strip since it is subjected to a heating process.

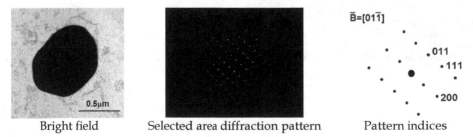

Bright field   Selected area diffraction pattern   Pattern indices

Fig. 16. Identification of an existing large particle observed in the as-received strip (Kim & Lee, 2002)

The remainder of the discussion explores several anomalous phenomena observed during the annealing of the alumina DS copper alloy.

4.1 Unique recrystallized microstructure

The recrystallized microstructure of the boron-added alumina DS copper alloy strip was characterized by the following features:

- Recrystallization only in the centre region
- Plate-like morphology of recrystallized grains
- Very large recrystallized grains

Figures 5 and 12 show that recrystallization occurred only in the centre region of the strips. In order to observe how the microstructure evolved, both the as-received strips and the cold-rolled strips were quickly annealed. Figure 17 shows optical micrographs of the as-received strip annealed at 923 K for 10 s and 15 min. Recrystallized grains emerged along lines originating exclusively from the centre region. Detailed TEM observations revealed that large bands were present in the deformed state in the centre region and appeared to promote recrystallization. Figure 18 shows a large band found in the centre region of the as-received strip, along with its orientation. Figure 19 reveals that similar bands were present in the cold-rolled strips. The orientations of the large bands in the as-received and cold-rolled strips included cube, RD-rotated cube, copper, and ND-rotated copper. Among them, the ND-rotated copper orientation was similar to the recrystallization texture {112}<312> observed in the annealed strips. These pre-existing large bands survived the early stages of annealing as shown in Figures 20 and 21.

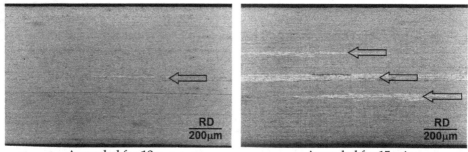

Annealed for 10 s   Annealed for 15 min

Fig. 17. Longitudinal section optical micrographs of the strips annealed at 923 K

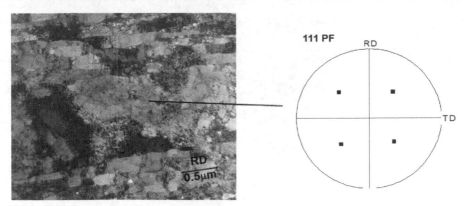

Fig. 18. Longitudinal section TEM micrograph showing a large band and its orientation in the centre region of the as-received strip (Kim & Lee, 2002)

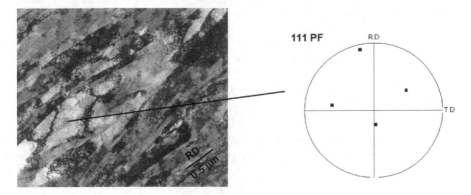

Fig. 19. Longitudinal section TEM micrograph showing a large band and its orientation in the centre region of the cold-rolled strip (Kim & Lee, 2002)

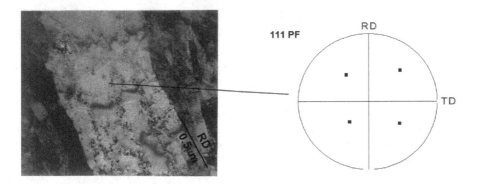

Fig. 20. Longitudinal section TEM micrograph showing a large band and its orientation in the centre region of the as-received strip annealed at 1123 K for 1 s

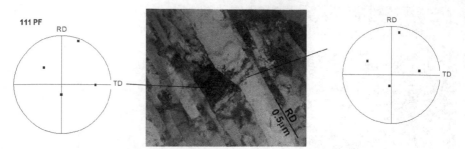

Fig. 21. Longitudinal section TEM micrograph showing a large band and its orientation in the centre region of the cold-rolled strip annealed at 1123 K for 1 s

Recrystallization can be divided into two consecutive processes: nucleation and growth. A nucleus must be some minimum size in order for further growth to occur, or else it will shrink and vanish. Subgrain coalescence is a requisite process to form these critical-sized nuclei. Band coalescence in the present material is unlikely when the band boundaries exhibit high-angle characteristics and their movement is hindered by the presence of dispersed particles. A more likely explanation is that the pre-existing large bands provide favourable nucleation sites for recrystallization. It appears that the large bands present in the centre region are the preferred recrystallization nucleation sites. It is possible that the large bands originated from large grains formed during the manufacturing process as similar grains have been observed in extruded alumina DS copper alloys in previous studies (Afshar & A. Simchi, 2008; H. Simchi & A. Simchi, 2009).

The plate-like morphology of the recrystallized grains in the alumina DS copper alloys can be related to the distribution of the dispersed particles. When recrystallized grains grow, the moving boundaries are pinned by the particles. The pinning pressure of the particles on the boundary movement is given by Equation 1 (Humphreys & Hatherly, 1995):

$$P_Z = 3F_V \gamma_b / d \tag{1}$$

where:

F_V is the volume fraction of the particles
d is the particle size
γ_b is the boundary energy

If particles are randomly distributed, the pinning pressure will be directionally isotropic. On the other hand, if the distribution of the particles is anisotropic, there will be an anisotropic pinning pressure on the boundaries. Figure 5 shows that the particles in the as-received strips were aligned along the rolling direction. The pinning pressure parallel to the rolling plane should be lower than that along the thickness direction. Therefore, the plate-like recrystallized grain shape can be mainly attributed to the planar distribution of the particles. The directional distribution of the particles might be driven by the rolling of the strip. A plate-like morphology of the recrystallized grains is often reported in the recrystallization behaviour of other dispersion-strengthened alloys (Klug et al., 1996; Chou, 1997), although other dispersion-strengthened alloys show equiaxed recrystallized grain structures (Miodownik et al., 1994; Miodownik et al., 1995).

Another unique recrystallization characteristic of the alumina DS copper alloy is that the recrystallized grains are very large. Early researchers (Singer & Gessinger, 1982; Mino et al., 1987; Kusunoki et al., 1990) reported that the very large recrystallized grains found in dispersion-strengthened alloys are formed through secondary recrystallization. They concluded that primary recrystallization occurred immediately before secondary recrystallization, or during plastic deformation – dynamic recrystallization. Later studies (Klug et al., 1996) suggested that primary recrystallization was responsible for the formation of large grains because microstructural changes are driven by stored energy acquired from plastic deformation. While plastically deformed alumina DS copper alloy possesses a sufficient driving force for recrystallization, a barrier to recrystallization exists due to the particle pinning effect. Microstructural inhomogeneity, such as large bands, provides preferential nucleation sites, and a large nucleus at a large band can grow with a size advantage over the surrounding matrix. Therefore, the emergence of very large recrystallized grains is a result of preferential nucleation at pre-existing large bands. The annealing behaviour of alumina DS copper alloy might be regarded as secondary recrystallization since very large recrystallized grains are formed when they overcome the particle-pinning pressure. However, the microstructure of the alumina DS copper alloy suggests that the driving force for recrystallization is stored energy by plastic deformation. Thus, while the evolution of the annealed alumina DS copper alloy microstructure appears to be due to secondary recrystallization, the mechanism that forms the very large recrystallized grains is due to primary recrystallization.

4.2 Unique recrystallization texture

The recrystallization texture of the annealed alumina DS copper alloy can be approximated by {112}<312>. To our knowledge, this texture has not been reported for other copper alloys. The recrystallization texture is determined by the orientations of the new grains and their growth rates. The present study discussed the role of these two factors and how they determine the unique recrystallization texture of the alumina DS copper alloy.

4.2.1 Selective nucleation

As discussed previously, pre-existing large bands provided favourable nucleation sites for recrystallization. Pre-existing large particles could introduce particle deformation zones that act as nucleation sites. Figure 22 shows the recrystallizing grains formed around the particles and their orientations observed in the as-received strips after rapid annealing. The orientation of grain A was similar to that of the deformed matrix, and multiple twinning could cause grains B and C to generate different orientations. It is known that PSN usually gives rise to weak recrystallization textures (Humphreys & Hatherly, 1995). Band coalescence is unlikely but possible when the pinning of the boundary movement is relaxed. Figure 23 shows that the band growth took place by coalescence of similarly oriented bands. Various orientations could be generated from new grains resembling the matrix orientations through PSN and band coalescence, as well as by subsequent twinning. Since no specific grain orientations dominated as the new grains evolved, the well-developed strong recrystallization texture {112}<312> could not be caused by new grain evolution.

4.2.2 Selective growth

According to the theory of selective growth, the recrystallization texture is determined by the relative growth rates of the boundaries. The velocity of the moving boundary (V) is a function of the boundary mobility (M) and the driving pressure (P), given by:

$$V=MP \tag{2}$$

P can be expressed as follows:

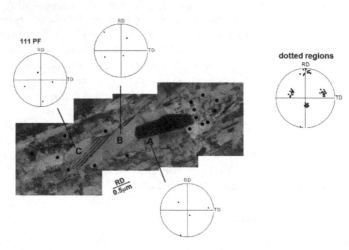

Fig. 22. Longitudinal section TEM micrograph showing individual grains around a particle and their orientations in the as-received strips annealed at 923 K for 10 s

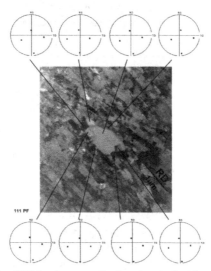

Fig. 23. Longitudinal section TEM micrograph showing individual grains and their orientations in the cold-rolled strips annealed at 1123 K for 3 s (Kim & Lee, 2002)

$$P = P_D - P_C = P_D - 2\gamma_b / R \tag{3}$$

where:

P_D is the stored energy,
P_C is the opposing pressure from the boundary curvature
γ_b is the boundary energy
R is the radius of the grain.

In particle-strengthened alloys, the Zener pinning pressure (P_Z) arises from the particles, and P can be expressed as follows (Humphreys & Hatherly, 1995):

$$P = P_D - P_C - P_Z = P_D - 2\gamma_b / R - 3F_V \gamma_b / d \tag{4}$$

where:

F_V is the volume fraction of the particles
d is the particle size.

Recrystallizing grains will grow only when P is positive. P increases with increasing grain size and decreasing boundary energy. The low-angle boundaries and twin boundaries have a lower boundary energy than the high-angle boundaries. Based on Equation 4, only large grains with low-angle boundaries or twin boundaries can overcome the pinning pressure. High-angle boundaries can be stagnant, even though they have higher mobility than low-angle boundaries. The recrystallization texture {112}<312> is defined as ND-rotated copper, which is occasionally found in large bands in the deformed state, as shown in Figures 19 and 21. Recrystallizing grains with {112}<312> orientations have a chance to face the surrounding deformed matrix with low-angle boundaries because {112}<312> orientations deviate slightly from the deformation texture. Furthermore, {112}<312> orientations have a twinning relationship between the two equivalent orientations among them. Figure 24 shows the orientations of two adjacent recrystallized grains observed in the cold-rolled and annealed strips. The boundary shape and orientation relationship indicated that the grain boundary of the two adjacent recrystallized grains was a twin boundary. Therefore, the unique recrystallization texture was determined by the preferential growth of large recrystallizing grains with low-angle boundaries or twin boundaries, even though those boundaries had low mobility.

 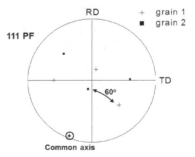

Fig. 24. Longitudinal section TEM micrograph and (111) pole figure showing two adjacent recrystallized grains of the cold-rolled strips annealed at 1123 K for 1 hr

4.3 Dependency of recrystallization on prior rolling conditions

As described in Section 3.2, the response to annealing of the alumina DS copper alloy is influenced by prior rolling conditions. The annealing behaviour of the cold-rolled strip is characterized by recrystallization, whereas recovery by band growth occurs in the hot-rolled strip. Similar results have been reported for other dispersion-strengthened alloys (Petrovic & Ebert, 1972; Singer & Gessinger, 1982). As recovery and recrystallization are competitive processes, dynamic recovery during hot rolling could reduce the potential energy in the alloy. This argument would also apply when comparing results between the cold-rolled and hot-rolled strips; since the hardness of the hot-rolled strip is lower, the recovery process during hot-rolling is governed by normal band growth. In our tests, after annealing, a continuous band growth occurred in the hot-rolled strip, which became a coarse band structure with high-angle boundary characteristics (see Table 7).

Subjecting the as-received strips to hot rolling gave rise to band growth and increased the hardness. Plastic deformation during hot rolling could increase the dislocation density, increasing the hardness. Therefore, a reduction in potential energy may not occur during hot rolling. It is not clear at this time why the hot-rolled strips became resistant to recrystallization. One explanation would be the homogeneity of the microstructural evolution. Microstructural inhomogeneity often occurs during plastic deformation, and these regions are frequently sites of initial recrystallization. The deformation becomes more homogeneous as the deformation temperature increases (Humphreys & Hatherly, 1995). A reduction in microstructural inhomogeneity during hot rolling could be responsible for the suppression of discontinuous recrystallization.

An alternative explanation is based on the assumption that coarse particles are sheared into finer particles during hot rolling (Kim & Lee, 2002). The shear strength of the particle might decrease with increasing temperature. Particle shearing could result in a decrease in interparticle spacing, which in turn could give rise to the higher hardness and the corresponding difficulty in recrystallization.

5. Conclusions

The recrystallization behaviour of boron-added alumina DS copper alloy strips was studied. The results may be summarized as follows.

Recrystallization occurred only in the centre region of the strips. Pre-existing large bands provided a favourable nucleation site for very large recrystallized grains.

The morphology of the recrystallized grains was plate-like due to the planar alignment of the dispersed particles.

The recrystallization texture was indexed to {112}<312>. Preferential growth of the large recrystallizing grains against the particle pinning appeared to determine this unique recrystallization texture.

The hot-rolled strip underwent recovery accompanied by continuous band growth, but without recrystallization.

6. References

Afshar, A. & Simchi, A. (2008). Abnormal Grain Growth in Alumina Dispersion-Strengthened Copper Produced by An Internal Oxidation Process. *Scripta Materialia*, Vol.58, No.11, (June 2008), pp.966-969, ISSN 1359-6462

Chou, T.S. (1997). Recrystallization Behavior and Grain Structure in Mechanically Alloyed Oxide Dispersion Strengthened MA956 Steel. *Materials Science and Engineering A*, Vol.223, No.1-2, (February 1997), pp. 78-90, ISSN 0921-5093

Gallagher, D.E.; Hoyt, E.W. & Kirby, R.E. (1988). Surface Segregation of Boron in Dispersion-Strengthened Copper. *Journal of Materials Science*, Vol.27, No.21 (November 1992), pp. 5926-5930, ISSN 0022-2461

Gessinger, G.H. (1976). Mechanical Alloying of IN-738. *Metallurgial Transactions A*, Vol.7, No.8, (August 1976), pp. 1203-1209, ISSN 0360-2133

Hirsch, J. & Lücke, K. (1988). Mechanism of Deformation and Development of Rolling Textures in Polycrystalline FCC Metals-I. Description of Rolling Texture Development in Homogeneous CuZn Alloys. *Acta Metallurgica*, Vol.36, No.11, (November 1988), pp. 2863-2882, ISSN 1359-6454

Humphreys, F.J. & Hatherly, M. (1995). *Recrystallization and Related Annealing Phenomena*, Pergamon, ISBN 0-08-041884-8, Oxford, United Kingdom

Kim, S.-H. & Lee, D.N. (2001). Recrystallization of Alumina Dispersion Strengthened Copper Strips. *Materials Science and Engineering A*, Vol.313, No.1-2, (August 2001), pp. 24-33, ISSN 0921-5093

Kim, S.-H. & Lee, D.N. (2002). Annealing Behavior of Alumina Dispersion-Strengthened Copper Strips Rolled Under Different Conditions. *Metallurgical and Materials Transactions A*, Vol.33, No.6 (June 2002), pp. 1605-1616, ISSN 1073-5623

Klug, R.C.; Krauss, G. & Matlock, D.K. (1996). Recrystallization in Oxide-Dispersion Strengthened Mechanically Alloyed Sheet Steel. *Metallurgical and Materials Transactions A*, Vol.27, No.7, (July 1996), pp. 1945-1960, ISSN 1073-5623

Kusunoki, K.; Sumino, K.; Kawasaki, Y. & Yamazaki, M. (1990). Effects of the Amount of γ' and Oxide Content on the Secondary Recrystallization Temperature of Nickel-Base Superalloys. *Metallurgical Transactions A*, Vol.21, No.2, (February 1990), pp.547-555, ISSN 0360-2133

Matthies, S.; Vinel, G.W. & Helming, K. (1987). *Standard Distribution in Texture Analysis*, Akademie-Verlag, ISBN 3-05-500249-0, Berlin, Germany

Mino, K.; Nakagawa, Y.G. & Ohtomo, A. (1987). Abnormal Grain Growth Behavior of An Oxide Dispersion Strengthened Superalloy. *Metallurgical Transactions A*, Vol.18, No.6, (June 1987), pp. 777-784, ISSN 0360-2133

Miodownik, M.A.; Martin. J.W. & Little. E.A. (1994). Secondary Recrystallization of Two Oxide Dispersion Strengthened Ferritic Superalloys : MA956 and MA 957. *Materials Science and Technology*, Vol.10, No.2, (February 1994), pp. 102-109, ISSN 0267-0836

Miodownik, M.A.; Humphreys, A.O. & Martin, J.W. (1995). Growth of Secondary Recrystallized Grains during Zone Annealing of Oxide Dispersion Strengthened Alloys. *Materials Science and Technology*, Vol.11., No.5, (May 1995), pp. 450-454, ISSN 0267-0836

Nadkarni, A. (1984). Dispersion Strengthened Copper Properties and Applications, In : *High Conductivity Copper and Aluminum Alloys*, E. Ling, P.W. Taubenblat, (Ed.), 77-101, TMS-ASME, Warrendale, PA

Petrovic, J.J. & Ebert, L.J. (1972). Electron Microscopy Examination of Primary Recrystallization in TD-Nickel. *Metallurgical Transactions*, Vol.3, No.5, (May 1972), pp.1123-1129, ISSN 0360-2133

Petrovic, J.J. & Ebert, L.J. (1972). Abnormal Grain Growth in TD-Nickel. *Metallurgical Transactions*, Vol.3, No.5, (May 1972), pp.1131-1136, ISSN 0360-2133

Preston, O. & Grant, N.J. (1961). Dispersion Strengthening of Copper by Internal Oxidation. *Transactions of the Metallurgical Society of AIME*, Vol.221, (February 1961), pp.164-173

Randle, V. (1993). *The Measurement of Grain Boundary Geometry*, Institute of Physics Publishing, ISBN 0-7503-0235-6, London, United Kingdom

Simchi, H. & Simchi, A. (2009). Tensile and Fatigue Fracture of Nanometric Alumina Reinforced Copper with Bimodal Grain Size Distribution. *Materials Science and Engineering A*, Vol.507, No.1-2, (May 2009), pp. 200-206, ISSN 0921-5093

Singer, R.F. & Gessinger, G.H. (1982). The Influence of Hot Working on the Subsequent Recyrstallization of a Dispersion Strengthened Superalloy-MA 6000. *Metallurgical Transactions A*, Vol.13, No.8, (August 1982), pp. 1463-1470, ISSN 0360-2133

Stephens, J.J. & Nix, W.D. (1985). The Effect of Grain Morphology on Longitudinal Creep Properties of INCONEL MA 754 at Elevated Temperature. *Metallurgical Transactions A*, Vol.16, No.7, (July 1985), pp. 1307-1324, ISSN 0360-2133

Sumino, Y.; Watanabe, H. & Yoshida, N. (2009). The Microstructural Evolution of Precipitate Strengthened Copper Alloys by Varing Temperature Irradiation. *Journal of Nuclear Materials*, Vol.386-388, No.C, (April 2009), pp. 654-657, ISSN 0022-3115

Young, C.T.; Steele, J.H. & Lytton, J.L. (1973). Characterization of Bicrystals Using Kikuchi Patterns, *Machine Design*, Vol.4, No.9, (September 1973), pp. 2081-2089, ISSN 0024-9114

2

Development of Texture and Substructure Inhomogeneity by Recrystallization of Rolled Zr-Based Alloys

Yuriy Perlovich and Margarita Isaenkova
National Research Nuclear University "MEPhI"
Russia

1. Introduction

Recrystallization of α-Zr is of the great interest both as a rather complicated scientific phenomenon and as a process of the practical importance for applications in the nuclear industry. Meanwhile in the most known monographs on Zr and Zr-based commercial alloys (Douglass, 1971; Tenckhoff, 1988; Zaymovskiy et al., 1994) the recrystallization of α-Zr is considered on the basis of experimental data, obtained more, than 40 years ago. These data urgently require corrections with taking into account the up-to-day theoretical conceptions and the continuous progress in experimental technique. Zr-based alloys are characterized by α↔β phase transformations within the technologically important temperature interval 610°-850°C and by operation of various mechanisms of α-Zr plastic deformation, including slip by basal, prismatic and pyramidal planes as well as twinning in several systems. These features are responsible for very complicated distribution of strain hardening and the corresponding tendency to recrystallization in products from Zr-based alloys. The given chapter makes up some gaps in our knowledge concerning different aspects of recrystallization as applied to α-Zr.

2. Regularities of recrystallization in sheets and tubes of Zr-based alloys with multicomponent rolling textures

The most widespread data on recrystallization regularities in α-Zr pertain to sheets and tubes with the final stable rolling texture of α-Zr $(0001)\pm20°\text{-}40°$ ND-TD $<10\bar{1}0>$ (Douglass, 1971), where ND – normal direction and TD – transverse direction. Meanwhile later the new detailed data were obtained concerning texture development in α-Zr under rolling. In particular, it was established that by rolling of a textureless slab the intermediate texture $(0001) \pm15°\text{-}25°$ ND-RD $<11\bar{2}L>$ forms, where RD – rolling direction, and keeps its stability up to the deformation degree of 70%, whereupon it converts to the final stable texture (Isaenkova & Perlovich, 1987a, 1987b). Main components of these textures in the order of formation were denoted as T1 and T2. Besides, components $(0001)<10\bar{1}0>$ (T0) and $\{11\bar{2}0\}<10\bar{1}0>$ (T3) are present often in textures of rolled sheets and, especially, tubes. All these texture components form owing to activity of concrete combinations of plastic deformation mechanisms, have their characteristic strain hardening and therefore show different tendencies to recrystallization. In the real case of a multicomponent texture the development of recrystallization must be additionally complicated by interaction between different components in regions of their contact. Indeed, according to (Isaenkova et al., 1988; Perlovich et al., 1989), resulting changes of α-Zr rolling textures in the course of recrystallization can not be reduced to 30°-rotation around basal normals and require for more complex description. In order to investigate this question in more details, the following work was undertaken.

2.1 Materials and methods

Recrystallization was investigated in sheet samples of alloys Zr-2,5%Nb, Zr-2,3%Cr and pure Zr as well as in tube samples of the alloy Zr-2,5%Nb. Sheets were produced by longitudinal or transverse cold rolling up to deformation degrees in the range from 40% to 90% in such a way as to form the following textures: T1, T1+T2, T2, T1+T2+T3. The weak component T0 was present everywhere. Perfection parameters and mutual relationship of different components varied. The channel tube was cold-rolled by 50%-thinning of its wall. All samples were annealed in dynamic vacuum at temperatures 500°-600°C during 1-5 h.

The main used method was X-ray diffractometric texture analysis. Direct pole figures (PF) (0001), $\{11\bar{2}0\}$ and $\{10\bar{1}2\}$ were measured by the standard procedure (Borodkina & Spector, 1981). To reveal PF regions, where texture changes by recrystallization are predominantly localized, diagrams of PFs subtraction (SD) were calculated and constructed. SD involves contours of pole density equal changes, having been drawn by comparison of recrystallization and rolling textures. In addition, PF sections of interest were constructed to follow redistribution of basal and prismatic normals in the course of recrystallization.

2.2 Recrystallization in sheets

Analysis of PFs $\{11\bar{2}0\}$ shows that texture changes by recrystallization can be described as rotation around the motionless basal axis only in the case of the rolling texture consisting largely of the component T2. However, the angle of such rotation varies: e.g. in the case of cold rolling by 60% for pure Zr recrystallized at 500°C this angle is equal to 30°, while for the alloy Zr-2,5%Nb recrystallized at 580°C – only 20°. When the rolling texture consists predominantly of the component T1, recrystallization does not involve lattice rotation around basal normals, - this is confirmed by invariance of PF$\{10\bar{1}2\}$.

Main results concern reorientation of basal normals by α-Zr recrystallization, i.e. changes of PF(0001). Superposition of SD and PF(0001) is shown in Fig. 1 for the sheet alloy Zr-2,5%Nb rolled up to deformation degrees 40, 60 and 80%, corresponding to formation of different textures: T1, T1+T2, T2. The densest cross-hatching indicates zones, where texture are localized predominantly. These zones are situated at slopes of initial texture maxima, increasing scattering of the recrystallization texture. Hence, the model of inhomogeneous strain hardening, proposed in (Perlovich, 1994) for textured BCC-metals, is true for HCP α-phase also. But while taking into account the multicomponent character of observed rolling textures, regularities of recrystallization should be more complicated.

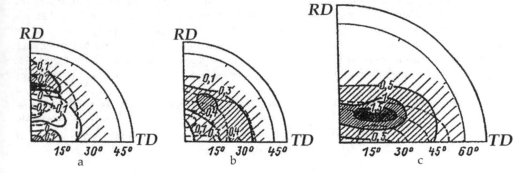

Fig. 1. Superposition of partial PF(0001) (dotted contours) for rolled sheets of the alloy Zr-2,5%Nb by deformation degrees 40% (a), 60% (b), 80% (c) and SD (solid contours), characterizing texture changes by recrystallization. Cross-hatching density reflects an increase of pole density by recrystallization.

For rolling textures T1+T2, where I(T1)>I(T2), it was established that the relative increase of main different components in the recrystallization texture depends on the initial relationship of these components in the rolling texture. In particular:

$$\Delta I(T2)/\Delta I(T1)=f_1[I(T2)/I(T1)], \tag{1}$$

$$\Delta I(T2)/\Delta I(T0)=f_2[I(T2)/I(T0)]. \tag{2}$$

These dependences are drawn in Fig. 2, where linearity of f_1 and nonlinearity of f_2 are evident. The nearer are initial intensities of components T2 and T1, the greater is difference between growth rates of new grains with corresponding orientations. As an area of the

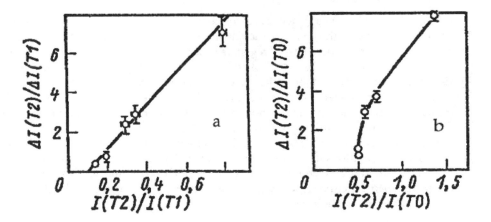

Fig. 2. Relative changing of texture components T2 versus T1 (a) and T2 versus T0 (b) in the recrystallization texture of sheet samples depending on the initial relationship of these components in the rolling texture.

contact surface between deformed grains of these components increases, conditions for growth of grains with the orientation T2 become more favorable. This signifies that T2-grains are growing into T1-grains and absorb them. The weaker dependence connects growths of components T2 and T0 in the recrystallization texture, since the difference between their strain hardening is less than in the case of components T2 and T1. Recrystallization of samples, showing predominance of the component T1, involves essential redistribution of basal normals even to the point of main texture component changing; then, depending on the concrete relationship of components in the rolling texture, T1 can give way to T0 (Fig. 3-a) or T2 (Fig. 3-b). Composition of the alloy influences mainly temperature parameters of the recrystallization process in α-Zr and seems to be of secondary importance for orientation regularities.

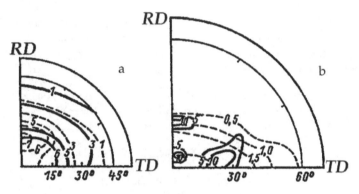

Fig. 3. Partial PF (0001) for sheet alloys Zr-2,5%Nb (a) and Zr-2,3%Cr (b) in rolled (dotted contours) and recrystallized (solid contours) states.

2.3 Recrystallization in tubes

The rolling texture of investigated tubes exhibits predominance of the component $\{11\bar{2}0\}<10\bar{1}0>$ and contains the intensive axial component with the axis $<10\bar{1}0>$ (Fig. 4). Distributions of X-ray reflection (0004) registered intensity in the PF section R-T (radial direction – tangential direction) both for rolled and annealed samples are presented in Fig. 5-a. Redistribution of registered intensity in consequence of tube annealing at 500°C is connected with a general increase of the intensity level by recovery (compare curves 1 and 2 in Fig. 5) and testifies about inhomogeneous recovery in grains with different orientations, resulting from their previous inhomogeneous strain hardening. The most active recovery is localized in grains with basal normals deflected from R-direction by 50°-60° and 90°. Hence, grains with such orientations have the greatest strain hardening and should show the maximal tendency to recrystallization. The curve 2 characterizes the true texture of the rolled tube better, than the curve 1, whose appearance on the inhomogeneous distribution of lattice defects in grains with different orientations.

As the annealing temperature increases, recrystallization process becomes more active and a sharp decrease of pole density near T-direction occurs, testifying about reorientation of basal normals. This effect is especially strong by passing from curve 4 to curve 5 (Fig. 5-a). An increase of the annealing temperature results in shifting of the maximum in the distribution of basal normals to R-direction: after annealing at 550°C is at angular distance of 45° from R-direction, while after annealing at 600°C – at a distance of 60°. The maximum corresponds to grains, which by recrystallization are in favorable conditions for growing, though their initial strain hardening is not maximal. These grains respond to the compromise variant: their strain hardening is so high, that the accumulated energy of lattice distortion ensures their sufficiently quick growth under recrystallization annealing, and at the same time the volume fraction of these grains is sufficiently large for absorption of a significant part of the deformed matrix.

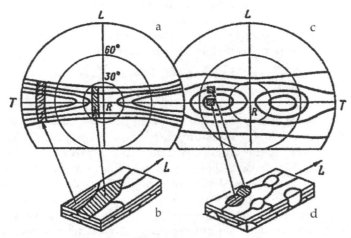

Fig. 4. PF(0001) (a, c) schematic images of microstructure (b, d) for the tube of the alloy Zr-2,5%Nb in rolled (a, b) and annealed (c, d) states. The microstructure (d) corresponds to the initial stage of recrystallization.

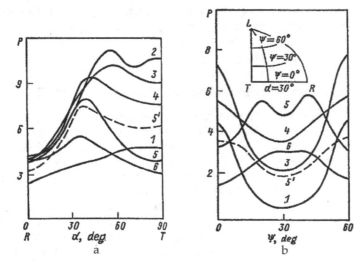

Fig. 5. Intensity distribution in the section R-T of PF(0001) (a) and in the section R-L of PF{11$\bar{2}$0} (b) for tubes of the alloy Zr-2,5%Nb: 1 – initial rolled state; 2-6 – annealing regimes: 2 - 500°C, 3 h; 3 - 530°C, 3 h; 4 - 550°C, 3 h; 5 - 580°C, 3 h; 5′ - 580°C, 1 h; 6 - 600°C, 3h.

As a result of tube annealing the new maxima arise on the meridional section of PF$\{11\bar{2}0\}$ (Fig. 5-b), testifying that, along with redistribution of basal normals in investigated tubes, reorientation of prismatic normals takes place also, leading to development of the typical for α-Zr recrystallization texture (0001) ± α R-T <$21\bar{3}0$ – $11\bar{2}0$ >. The situation of new maxima in the distribution of prismatic normals corresponds to slopes of initial maxima in the rolling texture, where strain hardening achieves increased values (Perlovich, 1994). The angle of misorientation between deformed and recrystallized grains by the common basal axis depends on the annealing temperature and increases from 20° to 30° when passing from annealing at 580°C to annealing at 600°C (compare curves 5 and 6 in Fig. 5-b). According to the data, presented in Fig. 5-a and 5-b, reorientation of basal normals is somewhat ahead of reorientation of prismatic normals both by time and temperature.

In order to explain the observed development of recrystallization in tubes, the model was proposed, suggesting operation of two different recrystallization mechanisms: (1) growth of new grains with intermediate orientations at high-angle boundaries between regions, having different local textures and originating from different initial grains; (2) growth of new grains with orientations, corresponding to zones of increased strain hardening within rolled initial grains. After rolling up to high deformation degrees, a polycrystal consists of thin plate-like grains, which have their own local textures and contain only low-angle subboundaries. Such structure is shown in Fig. 4-b. Then the first mechanism causes formation of new grains with intermediate orientations of basal normals, as it is shown in Fig. 4-c,d, while the second mechanism results in lattice rotation about basal normals owing to gradual growing of nuclei in zones of increased strain hardening within plate-like grains. Absorption of the deformed matrix by nuclei of new grains, growing along T- and R-directions, is controlled by structure anisotropy of a rolled tube.

Thus, it was shown that by recrystallization of α-Zr along with previously described lattice rotation around basal normals a significant redistribution of these normals takes place both in sheets and tubes. Contrary to the widespread idea, the recrystallization texture of α-Zr varies in a wide range depending on the relationship of main components in the rolling texture.

3. Substructure inhomogeneity of recrystallized sheets from Zr-based alloys

The substructure inhomogeneity is a generally recognized feature of deformed metal materials, controlling the process of their recrystallization. Since the plastic deformation usually is accompanied by arising of the crystallographic texture, developments of the substructure inhomogeneity and the deformation texture prove to be mutually interconnected. It was shown by means of X-ray diffractometric methods (Perlovich et al., 1997; 2000; Perlovich & Isaenkova, 2002), that an actual spectrum of substructure conditions in rolled metals and alloys is extremely wide and that the optimal criterion for systematization of observed substructure inhomogeneities is the grain orientation. The main principle of substructure inhomogeneity is the following: by passing from texture maxima to texture minima, grains (subgrains, blocks etc.) become finer and the lattice distortion increases. A question arises whether the recrystallization removes this inhomogeneity or recrystallized material partially retains it, contrary to the widespread viewpoint concerning its negligible scale.

The above question is very actual as applied to commercial Zr-based alloys in connection with their usage as responsible construction materials in nuclear reactors. The final heat treatment of products of Zr alloys is aimed to attain the stability of their structure and to remove residual stresses of all kinds. The recrystallization is usually believed to satisfy these requirements, since experimental evidences of its efficiency are restricted by those, accessible by standard methods of structure characterization. But standard X-ray methods are selective, i.e. obtained data relate only to grains with some definite orientation, corresponding to their reflecting position by the used measurement geometry. Therefore, these data can be considered as sufficient only under a supposition that recrystallization practically removes the substructure inhomogeneity of products. An aim of the given study is to demonstrate the real substructure inhomogeneity of recrystallized products of Zr-1%Nb and Zr-2.5%Nb alloys using special methods of modern X-ray diffractometry.

3.1 The X-ray method of generalized pole figures

Recent development of X-ray diffractometric technique allowed to elaborate a new method of the fullest description of textured metal materials with taking into account their substructure inhomogeneity. The method involves repeated recording of X-ray line profiles by the geometry of texture measurements, so that, as opposed to the standard description of the substructure by parameters of the X-ray reflection (hkl) from crystallographic planes {hkl} of the single orientation, now the substructure condition of the sample can be characterized by the multitude of line profiles, corresponding to planes {hkl} within grains of different orientations.

The treatment of measured data includes correction for the defocalization effect, approximation of X-ray line profiles with pseudo-Voight functions, calculation of their integral intensity I, physical half-width β and peak position 2θ, construction of distributions $I(\psi,\varphi)$, $\beta(\psi,\varphi)$, $2\theta(\psi,\varphi)$ in the stereographic projection of the sample, where (ψ,φ) – coordinates of reflecting planes {hkl}. These distributions, named Generalized Pole Figures (GPF), characterize substructure conditions along axes <hkl> by all their space orientations. In particular, normalized GPF $I_{hkl}(\psi,\varphi)$ is the usual texture pole figure PF{hkl}, GPF $\beta_{hkl}(\psi,\varphi)$ exhibits the combined effect of coherent block size D_{hkl} and lattice distortion $\Delta d/d_{hkl}$, GPF $2\theta(\psi,\varphi)$ describes the anisotropic elastic deformation $\varepsilon_{hkl}(\psi,\varphi)$ of grains along axes <hkl> due to action of residual microstresses.

The measured GPF $2\theta_{hkl}(\psi,\varphi)$ can be recalculated into GPF $d_{hkl}(\psi,\varphi)$ and further – into GPF $\varepsilon_{hkl}(\psi,\varphi)$, where $\varepsilon_{hkl}(\psi,\varphi) = [d(\psi,\varphi)-d_{av}]/d_{av}$ and d_{av} – the averaged weighted value of interplanar spacing d_{hkl}. Depending on the sign of ε_{hkl}, elastic extension or elastic contraction takes place along axis <hkl> with coordinates (ψ,φ), so that GPF $\varepsilon_{hkl}(\psi,\varphi)$ allows to reconstruct a deformation tensor for grains of main texture components.

Diagrams of the correlation between different GPF are very useful by the analysis of regularities, controlling formation of the inhomogeneous substructure in real metal materials by their technological treatment. These diagrams are constructed in coordinates (β_{hkl}, I_{hkl}) or $(2\theta_{hkl}, I_{hkl})$, so that each their point corresponds to some point (ψ_i, φ_j) in the stereographic projection of the sample and in GPF of reference. When taking into account, that each crystallite of α-Zr, having HCP lattice, has only one axis <0001>, volume fractions of grains with different physical broadening β and peak positions 2θ of X-ray reflections from basal planes (0001), as well as with different values of derivative substructure characteristics along c-axis, can be determined. With this aim for all points (ψ,φ) of PF (0001) values of pole density are recalculated into weight coefficients to be used by the statistical treatment of GPF for the parameter of interest.

3.2 Experimental details and results

Substructure features of recrystallized plates of Zr-1%Nb and Zr-2.5%Nb alloys were studied. Plates were obtained by plain and transverse cold rolling by $\varepsilon \cong 55\%$ of bars, which were cut out from the annealed slab 2 mm in thickness. The direction of cold rolling coincided either with RD of the initial slab or with its TD. All plates were annealed in the evacuated vessel at 580°C during 3h, so that in both alloys the recrystallization of the dominant α-Zr phase took place. The full cycle of X-ray measurements as applied to all plates was carried out twice, that is after rolling and after annealing. The X-ray study was preceded by etching of samples, aimed to remove the surface layer ~40 µm in thickness.

By X-ray studies the texture diffractometer SIEMENS D500/TX with a position sensitive detector was used. The profile of the same X-ray line was registered by each of 1009 successive positions of the sample in the course of texture measurement. For the data treatment both the supplied software and the original programs, elaborated by authors, were applied. All data, obtained for recrystallized samples, are considered in comparison with data for the same samples in the rolled state.

In Fig. 6 incomplete GPF are presented for studied samples both in rolled and recrystallized conditions. All GPF were constructed by measurements of the X-ray line $(0004)_{\alpha\text{-}Zr}$. In Fig. 7 distributions of volume fractions of grains, characterized by different half-widths β of the X-ray line (0004) and by different values of interplanar spacing d_{0001} are constructed for some studied samples. Fig. 8 shows correlation diagrams of GPF $\beta_{0004}(\psi,\varphi)$ and GPF $2\theta_{0004}(\psi,\varphi)$ with PF(0001) for as-rolled (o) and annealed (+) samples.

3.3 Features of obtained distributions

Consideration of obtained data allows to establish the following:

1. As a result of recrystallization, the texture of α-Zr in studied plates changes in accordance with principles, revealed in section 1, so that mutual ratios of main components in the recrystallization texture depend on these ratios in the rolling texture. The most sharp texture changes accompany recrystallization in the Zr-1%Nb plate, obtained by transverse rolling (Fig. 6): the texture with predominance of components $(0001)\pm15°\div20°$ND-RD <10.L>, stable by intermediate deformation degrees, transforms into the texture of the central type.

2. Though the significant part of α-Zr crystallites by recrystallization shows a drop of the physical broadening of X-ray line (0004) down to minimal measurable values (Fig. 7), corresponding to the coherent domain size above ~150 nm, the substructure inhomogeneity of all annealed samples is still rather essential and its general character remains the same, i.e. coherent domains becomes smaller and lattice distortions increases by passing from central regions of texture maxima to their periphery. In GPF $\beta(\psi,\varphi)$ regions with the most perfect substructure are darkened, so that its perfection increases with the degree of darkening (Fig. 6).

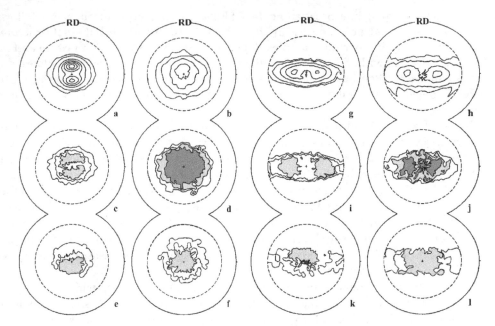

Fig. 6. PF(0001) (a, b, g, h), GPF β_{0004} (c, d, i, j) and GPF ε_c (e, f, k, l) for α-Zr: Zr-1%Nb, transversal rolling, deformed state - (a, c, e), annealed state - (b, d, f); Zr-2.5%Nb, plain rolling, deformed state - (g, i, k), annealed state - (h, g, l). Darkening: GPF β - the dark weakens from $\beta_1= 0{,}2°$ to $\beta_2= 0{,}8°$; GPF ε_c - the dark shows regions with $\varepsilon_c< 0$.

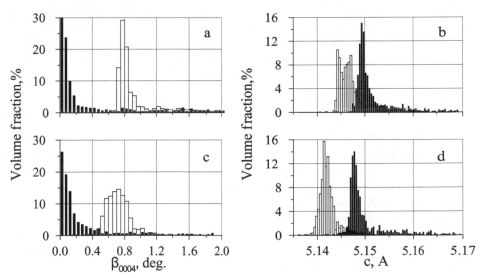

Fig. 7. Volume fractions of α-Zr crystallites with different values of X-ray line broadening β_{0004} (a, c) and lattice parameter "c" (b, d): Zr-1%Nb, transversal rolling – (a, b); Zr-2.5%Nb, plain rolling – (c, d). White columns – deformed state, black columns – annealed state.

3. Contours of equal line broadening in GPF $\beta(\psi,\varphi)$ for recrystallized plates follow contours of equal pole density in their PF(0001), testifying that the substructure inhomogeneity of annealed samples depends on the recrystallization texture and is not connected with their initial rolling texture. Hence, recrystallized plates do not succeed their substructure inhomogeneity to deformed material, but development of this inhomogeneity accompanies formation of the recrystallization texture.

4. Recrystallization results in the decrease of the average elastic microstrain, but nevertheless there are regions in GPF $\varepsilon_{0001}(\psi,\varphi)$ for recrystallized samples, where $\Delta c/c_{av}$ attains rather high values, comparable with those for as-rolled samples. In the orientation space a redistribution of elastic contraction and elastic extension along basal axes takes place in the course of recrystallization (Fig. 6), resulting in the change of the mode of microstress equilibrium (Perlovich et al., 1998): crystallites of rolled α-Zr experience residual contraction and extension along axes <0001>, deflected predominantly in opposite directions from the plane ND-TD, whereas in recrystallized α-Zr the region of elastic contraction surrounds ND and regions of elastic extension are shifted to the plane TD-RD.

5. The increase of c-parameter by recrystallization (Fig. 7) is connected with annealing of lattice defects and with leaving of excessive Nb atoms from the α-Zr solid solution in accordance with the balanced phase diagram (Douglass, 1971). Therefore, the difference in average values of c-parameter between rolled and annealed samples proves to be greater for the alloy with the higher content of Nb (2.5%).

6. In diagrams of correlation between GPF $2\theta_{0004}(\psi,\varphi)$ and PF(0001) (Fig. 8) it is distinctly seen that as a result of recrystallization the distribution of interplanar spacing d_{0001} does not become more homogeneous, than it was in the same sample after rolling; but it acquires the more regular character. Quite evident submission of interplanar spacings in recrystallized samples to usual statistical regularities reflects spontaneity of thermally activated processes, whereas in as-rolled samples these regularities are suppressed by forces, controlling plastic deformation processes.

Obtained experimental evidences of the essential substructure inhomogeneity in recrystallized Zr-based alloys depending on their texture prompt a number of inferences, connecting different aspects of recrystallization:

-   The recrystallization texture includes a wide spectrum of grain orientations, so that a significant mutual misorientation of neighboring grains is probable.
-   Significant microstrains can arise by meeting of neighboring growing grains and, as a result, a local elastic deformation and an increase of the dislocation density take place within boundary regions.
-   Formation of specific dislocation arrangements near grain boundaries are accompanied by some local lattice rotations, so that by passing from the central part of a grain to its periphery the orientation changes, at least, by several degrees.
-   Texture minima, where the fraction is localized with the most distorted crystalline lattice and finest coherent blocks, correspond to regions near boundaries of recrystallized grains.

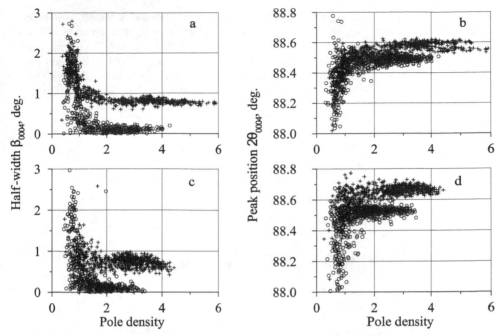

Fig. 8. Correlation diagrams for GPF β_{0004} and PF(0001) - (a, c), GPF $2\theta_{0004}$ and PF(0001) - (b, d): Zr-1%Nb, transversal rolling - (a, b); Zr-2.5%Nb, plain rolling – (c, d). Designation: + - deformed state, o - annealed state.

4. Texture of dynamic recrystallization

The question about an effect of dynamic recrystallization on the texture of rolling at increased temperatures emerges often by study of products from Zr-based alloys. This question is of the general interest and can be answered only by the rather attentive investigation of semi-products, produced at intermediate technological stages. An example of such investigation is presented below.

4.1 Materials, experimental approach and results

Sheets from Zr-2.5%Nb alloy, rolled at 750°C by two deformation routs down to thickness of ~4 mm, were studied. The used routes A and B differ in values of reduction per pass, diminishing in the order A→B, in numbers of successive passes and in presence of intermediate heating. The smaller are reductions per pass, the longer is the rolling procedure and the more significant is cooling of the billet (Perlovich et al., 2006). For compensation of this cooling the intermediate heating is introduced in route B. The alloy Zr-2.5%Nb contains usually two phases - the prevalent low-temperature HCP α-phase and the secondary high-temperature BCC β-phase. The temperature boundaries of (α+β)-region for Zr-2.5%Nb alloy are 610° and 830°C (Douglass, 1971), so that the studied sheets were rolled at the temperature of (α+β)-region, but local temperatures as well as the phase composition of concrete layers under rolling depended on their distance from the surface.

The layer-by-layer study of rolled sheets was carried out by X-ray diffractometric methods and mainly – by texture analysis. Texture analysis of α-Zr included measurement and construction of direct pole figures PF(0001) and PF{11$\bar{2}$0}. Typical PF(0001) and PF{11$\bar{2}$0} for surface, intermediate and central layers of the hot-rolled sheet are presented in Fig. 9. When passing from the surface layer of sheet to the central one, the rolling texture changes essentially. The following characteristic features of the rolling texture are considered: (1) the angular distance γ of texture maxima from normal direction (ND) in PF(0001); (2) the presence of additional texture maxima in PF{11$\bar{2}$0}, arising usually by recrystallization at the angular distance of 30° from maxima of the deformation texture. In PF{11$\bar{2}$0} (Fig. 9) these maxima are denoted by letters R and D, respectively. Layer-by-layer changes of angle γ are shown in figure 10-a; the layer-by-layer inhomogeneity of recrystallization is characterized by changes of ratio P_{recr}/P_{def} in Fig. 10-b, where P_{recr} and P_{def} – intensities of corresponding texture maxima.

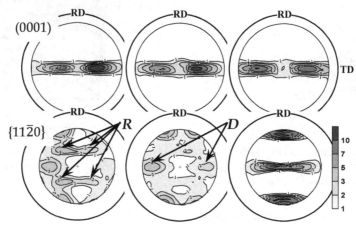

Fig. 9. Typical PF(0001) and PF{11$\bar{2}$0} for surface, intermediate and central layers of the hot-rolled sheet. Angular radius of constructed PF(0001) is equal to 80°, PF{11$\bar{2}$0} – 70°.

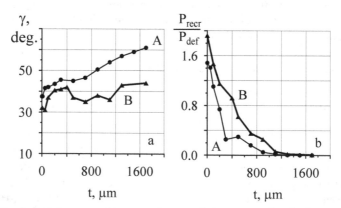

Fig. 10. Layer-by-layer changes of texture characteristics across the thickness of hot-rolled sheets: a – angular distance γ of texture maxima from ND in PF(0001); b - ratio P_{recr}/P_{def}, measured by PF{11$\bar{2}$0}.

4.2 Dynamic recrystallization under rolling

Development of recrystallization in α-Zr under rolling is seen in the surface layer up to 0.8 mm thick (Fig. 10-b). In PF{11$\bar{2}$0} of this layer (Fig. 9) maxima of rolling and recrystallization textures are present simultaneously, testifying that recrystallization of the sheet was only partial. Since texture maxima D and R are lying on different meridians, it is clear, that recrystallization included only grains with basal axes, closest to TD. Fig. 9 demonstrates visually the layer-by-layer inhomogeneity of α-Zr recrystallization in hot-rolled sheets. The angular shift of texture maxima in PF(0001) by passing from surface to central layers (Fig. 10-a) testifies in accordance with results of section 2, that recrystallization involves the reorientation of basal axes along with the known rotation of prismatic axes.

One more noteworthy difference between texture maxima D and R consists in their shape: maxima R are narrow and stretched along parallels of PF, likewise texture maxima in corresponding PF(0001), whereas maxima D are roundish and almost equiaxial. The shape of maxima R is a distinct evidence of dynamic recrystallization: anisotropic development of diffusion processes under hot rolling, including dislocation climb, results in stretching of texture maxima, as it is typical for rolling textures.

And vice versa, when the structure reforms spontaneously, only due to thermal activation, preferred directions in displacements of dislocations and dislocation boundaries are absent, so that texture maxima prove to be equiaxial, as it is usual for textures of static recrystallization. Then the round shape of maxima D in the rolling texture shows, that corresponding grains of α-Zr had time for polygonization in the course of cooling. A volume fraction of recrystallized grains decreases with distance from the surface, since stress relaxation in inner layers occurs by means of α→β→α phase transformations.

By route B the layer of dynamic recrystallization is thicker, than by route A (Fig. 10-b), because this process most probably develops in α-region of Zr-Nb phase diagram and the temperature boundary between α- and (α+β)-regions shifts deep into the sheet by passing from route A to route B in consequence of more intense cooling.

Thus, presented experimental results show, that by rolling of sheets from Zr-2.5%Nb alloy at the temperature 750°C a significant gradient in deformation conditions across the thickness of sheet takes place. The real temperature of concrete layers deviates from the nominal one due to the heat sink to rolls and local heating by deformation. A decrease of the deformation rate promotes development of dynamic recrystallization, suppresses the deformation-induced α→β phase transformation and weakens the unfavorable texture component, formed by rolling in β-phase.

5. Competition between recrystallization and phase transformations by welding of sheets from Zr-2,5%Nb alloy

The usual temperature of recrystallization annealing for α-Zr in cold-rolled products from Zr-based alloys is 580°C, that is close to the lower boundary of the (α+β)-region in the Zr-Nb phase diagram (610°C), where phase transformation (PT) α→β begins. Therefore under conditions of some heat treatments a competition is probable between recrystallization of α-Zr and PT α→β. In particular, such conditions take place by welding of cold-rolled sheets from Zr-based alloys in the thermal influence zone (TIZ) of the welding seam. When taking into account the regular difference between textures of sheets, experienced PT α→β without and after preliminary recrystallization (Cheadle & Ells, 1966), the inhomogeneity of recrystallization within TIZ of the welding seam was analyzed.

5.1 "Multiplication" of maxima in α-Zr texture by phase transformations α→β→α

The preliminary recrystallization of α-Zr influences the texture, which arise in the sheet as a result of PT α→β→α due to realization of the Burgers orientation relationship between α- and β-phases (Douglass, 1971):

$$(0001)_\alpha \parallel \{011\}_\beta, \ <11\bar{2}0>_\alpha \parallel <111>_\beta.$$

Multiplication of initial orientations in consequence of PT α→β→α was analyzed in (Cheadle & Ells, 1966). By absence of variant selection, 35 new orientations of the basal plane arise in addition to the initial one; by mutual coincidence of some orientations their total number decreases down to 24. Since in reality we deal with textured polycrystals of Zr-alloys instead of single crystals, after PT α→β→α the resulting distribution of basal axes in PF(0001) consists of overlapping texture maxima rather than of separate points. Therefore some new orientations of 24 above-mentioned ones, being close to each other, form common maxima. Thus, PT complicates an initial texture, multiplying its maxima in a definite way, though the resulting PF(0001) of the treated sample contains a lesser number of separate maxima, than the Burgers relationship predicts.

5.2 Studied samples and investigation technique

Samples for investigation were cut from sheets of the Zr-2,5%Nb alloy, cold-rolled up to the deformation degree of ~80%. In order to prepare semi-recrystallized samples, after cold rolling they were annealed at 550°C for 1 h. In order to induce PT α→β→α, both cold-rolled and annealed samples were subjected to the heat treatment in dynamic vacuum, including heating up to 950°C for 0,25 h, holding at this temperature during 0,5 h and subsequent cooling with an evacuated envelope in air. X-ray texture measurements were carried out by the standard method (Borodkina & Spector, 1981) using the diffractometer DRON-3M and Cu K_α radiation. For construction of complete PF, three mutually perpendicular sections of sheet or tube were investigated to obtain partial PF for their following sewing together (Perlovich & Isaenkova, 2002).

The investigated welded joints were produced by argon-arc welding of cold-rolled sheets of Zr-2,5%Nb alloy, using a non-expendable tungsten electrode with a motion velocity of 60 m/h. The welding direction (WD) was perpendicular to the RD. Texture inhomogeneity near the welding seam, connected with different heating conditions at neighboring regions, was studied layer-by-layer depending on the layer distance from the centre line of the seam.

5.3 Changes in α-Zr texture of cold-rolled and annealed sheets

PF(0001) for cold-rolled sheet is shown in Fig. 11-a, PF(0001) for the same sheet after heat treatment at 950°C – in Fig. 11-b, distributions of pole density along the radius ND-TD – in Fig. 11-c. Analogous PF and distributions for the sheet, which before PT experienced annealing at 550°C, - in Fig. 12-a,b,c. Comparison of PF(0001) in Fig. 11-a and Fig. 12-a shows, that partial recrystallization of the cold-rolled sheet causes extension of the pole density distribution to TD. Black circles in Fig. 11-b indicate ideal positions of texture maxima, arising as a result of PT α→β→α in PF(0001) of the cold-rolled sheet, whereas black circles in Fig. 12-b – calculated positions of PT-induced texture maxima in PF(0001) of the completely recrystallized sheet. In the considered real case of the semi-recrystallized sheet these black circles get only to those texture maxima, which are absent in PF(0001) of the cold-rolled sheet. This feature testifies, that some texture maxima in PF(0001) (Fig. 12-b) are produced by "multiplication" of maxima, belonging to the rolling texture, and some others – to the recrystallization texture.

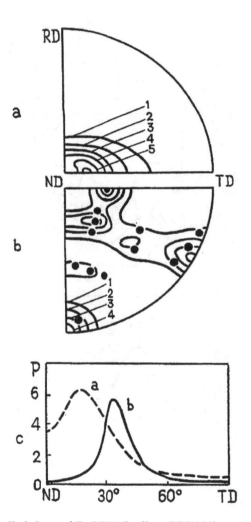

Fig. 11. PT in the cold-rolled sheet of Zr-2,5%Nb alloy, PF(0001): a – cold rolling; b – cold rolling + heat treatment 950°C/0,25 h; c – distributions of pole density along PF radius ND-TD.

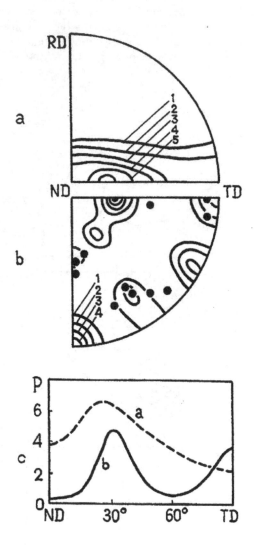

Fig. 12. PT in the semi-recrystallized sheet of Zr-2,5%Nb alloy, PF(0001): a – cold rolling + annealing 550°/1 h.; b - cold rolling + annealing 550°/1 h + heat treatment 950°/0,25 h; c - distributions of pole density along PF radius ND-TD.

In particular, as a result of PT $\alpha{\to}\beta{\to}\alpha$ in recrystallized grains, a new texture component $\{10\bar{1}0\}<11\bar{2}0>$ arises, having its maximum in PF(0001) at TD. Other maxima of the same set are present at the predicted regions also, confirming additionally the fact of preliminary recrystallization in the course of heating to the β-phase. At the same time, maxima, originating from deformed α-grains, by PT become noticeably weaker, than in the case of PT without preliminary recrystallization (compare Fig. 11-b and 12-b). The real situation in the deformed textured α-phase requires a description, similar to the case of a composite; therefore it would be correct to state a possibility of different outcomes from the competition between recrystallization and PT $\alpha{\to}\beta$ depending on deformation degree, grain orientation, heating rate, and so on. Inhomogeneous development of the considered processes, as well as their mutual competition, corresponds apparently to the most general case.

5.4 Recrystallization in the thermal influence zone by welding

Below some observations are presented concerning the competition between recrystallization and PT $\alpha{\to}\beta$ in the thermal influence zone (TIZ) with reference to the Zr-2,5%Nb alloy.

Arc welding is accompanied by a local heat treatment of the material in the vicinity of the welding seam. Parameters of a short-time thermal cycle, passing lengthwise TIZ parallel to the welding direction, are different for each longitudinal section of TIZ and depend on its distance from the central line of the seam. Layer-by-layer study of the texture within the TIZ gives an insight into the inhomogeneous structure developed in this zone by welding. In Fig. 13-a a schematic image of a welded joint is drawn; the melting zone is denoted by dense hatching and TIZ – by thin hatching. Three longitudinal sections are shown within the TIZ, and for each section an arrow indicates the corresponding PF(0001), obtained by X-ray diffractometric study just of this section. RD is brought into the centre of these PF in contrast with above-presented PF(0001) in Fig. 11 and 12.

Judging from PF in Fig. 13, different textures have formed in the shown sections of the TIZ depending on the distance from the seam. While in the most remote section the initial distribution of basal axes remains unchanged, the textures of the two closer sections contain new components produced by PT. Pole figures for these sections differ in relationship of PT components, originated from deformed (A) and recrystallized (B) α-grains. A quantitative treatment of obtained experimental data included the calculation of parameters, characterizing the relative contributions of both deformed and recrystallized components in the measured texture. In Fig. 13-b the results of such treatment for 18 successive sections of the TIZ are presented. The upper curve, constructed by PF$\{11\bar{2}0\}$, characterizes the relative fraction of recrystallized grains depending on the distance from the seam, irrespective of whether these grains experienced PT or did not. The following curve, constructed by PF(0001), characterizes the relative fraction of recrystallized grains in volume, covered by PT; it should be noted that in the general case this volume forms only a part of the investigated layer. The lower curve was obtained by subtraction of the (0001)-curve from the $\{11\bar{2}0\}$-curve. It characterizes the fraction, falling on grains, which proved to be recrystallized, but did not experience PT.

The presented curves testify unambiguously that, by sufficient increase of the heating temperature, all grains experience recrystallization prior to PT $\alpha{\to}\beta$. Thus, competition

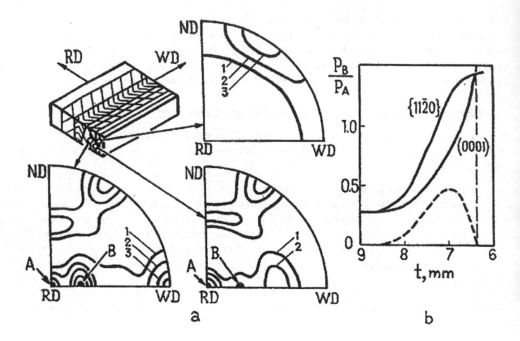

Fig. 13. PT in the thermal influence zone by arc welding: (a) the schematic image of welding seam and PF(0001) for its different section; (b) quantitative treatment of PF (see detailed explanation in the text).

between recrystallization and PT by a high rate of heating results in an absolute predominance of primary recrystallization (variant 1), though at some intermediate regimes of heat treatment two other variants are possible: recrystallization of deformed α-grains without their subsequent PT (variant 2) and PT of deformed α-grains without their preliminary recrystallization (variant 3).

Thus, the most important feature of the processes in the thermal influence zone is their inhomogeneous character, particularly by intermediate regimes of the heat treatment. While in some grains the variant 1 realizes, in other grains the variants 2 or 3 take place. The concrete variant, actual for the given grains, depends in some complicated manner on their orientation.

6. References

Borodkina M.M. & Spector E.N. (1981). *X-ray Analysis of Texture in Metals and Alloys*. Publishing House "Metallurgiya", Moscow, pp. 48-91 (in Russian).

Cheadle B.A. & Ells C.E. (1966) The effect of heat treatment on the texture of fabricated Zr-rich alloys. *Electroch. Techn.*, Vol. 4, No 7-8, pp. 329-336.

Douglass D.L. (1971). *The Metallurgy of Zirconium*. International Atomic Energy Agency, Vienna, pp. 60-76.

Isaenkova M. & Perlovich Yu. (1987a). Kinetics and mechanisms of texture formation in α-Zr by rolling. *Fizika Metallov i Metallovedenie*, Vol. 64, No 1, pp. 107-112 (in Russian).

Isaenkova M. & Perlovich Yu. (1987b). Reorientation of α-Zr crystallites by deformation. *Izvestiya Akademii Nauk SSSR. Metalli*, No 3, pp. 152-155 (in Russian).

Isaenkova M., Kapliy S., Perlovich Yu. & Shmelyova T. (1988). Features of changes of the Zirconium rolling texture by recrystallization. *Atomnaya Energiya*, Vol. 65, No 1, pp. 42-45 (in Russian).

Perlovich Yu., Isaenkova M., Shmelyova T., Nikulina A. & Zavyalov A. (1989). Texture changes in tubes of the alloy Zr-2,5%Nb by recrystallization. *Atomnaya Energiya*, Vol. 67, No 5, pp. 327-331 (in Russian).

Perlovich Yu. (1994). Development of strain hardening inhomogeneity during texture formation under rolling of bcc-metals. In: *Numerical Prediction of Deformation Processes and the Behavior of Real Materials, 15th Riso International Symposium on Material Science*, 5-9 September 1994, S.I. Andersen et al. Eds, Riso National Laboratory, Roskilde, Denmark, pp. 445-450.

Perlovich Yu., Bunge H.J. & Isaenkova M. (1997). Inhomogeneous distribution of residual deformation effects in textured BCC metals. *Textures & Microstructures*, Vol. 29, pp. 241-266.

Perlovich Yu., Bunge H.J., Isaenkova M. & Fesenko V. (1998) The Distribution of Elastic Deformation in Textured Materials as Revealed by Peak Position Figures. *Material Science Forum*, Vol. 273-275, pp. 655-666.

Perlovich Yu., Bunge H.J. & Isaenkova M. (2000) Structure inhomogeneity of rolled textured niobium. *Zeitschrifft fur Metallkunde*, Materials Research and Advanced Techniques, 2000, Vol. 91, No 2, p. 149-159.

Perlovich Yu. & Isaenkova M. (2002) Distribution of c- and a-dislocations in tubes of Zr alloys. *Metallurgical and materials transactions A*, Vol. 33A, No.3, pp. 867-874.

Perlovich Yu., Isaenkova M., Akhtonov S., Filippov V., Kropachev S. & Shtutca M. (2006) Interdependence of plastic deformation and phase transformations in Zr-2.5%Nb alloy under forging by different temperature-rate regimes. *Proceedings of the 9th International Conference on Material Forming ESAFORM 2006*, Glazgow, United Kingdom, April 2006, pp. 439-442.

Tenckhoff E. (1988) Deformation mechanisms, texture and anisotropy in Zirconium and Zircaloy. - ASTM, Special technical publication (STP 966), Philadelphia, 1988. - 77 p.

Zaymovskiy A.S., Nikulina A.V. & Reshetnikov N.G. (1994). *Zirconium Alloys in Nuclear Industry*. Energoatomizdat, Moscow, ISBN 5-283-03767-3, Russia, 256 p. (in Russian).

3

Crystal Growth:
Substructure and Recrystallization

Vadim Glebovsky
Institute of Solid State Physics, the Russian Academy of Sciences
Russia

1. Introduction

Single crystalline refractory transition metals (molybdenum, tungsten, niobium, and tantalum) exhibit a unique combination of properties, namely, high strength, plasticity, Young modulus, wear resistance, number, and low coefficient of linear expansion as well as high radiation resistance which is what makes single crystals of these metals of high purity the most suitable materials to be widely employed in science and engineering. Single crystals of high purity tungsten are well suited to production of the deflectors of the charged particles beams in the linear accelerators, the colliders and, also, to be successfully used as the target-converters for the sources of the positron beams. The attributes of single crystals of molybdenum alloys, compared to their polycrystalline counterparts, include more stable microstructures, lower creep rates, better compatibility with nuclear fuels and lower diffusion penetrability (Liu & Zee, 1996). On the other side, studies of the X-ray wave field in crystals or so called effects of the dynamic scattering theory are of high interest although the first observations of X-ray anomalous transmission are made more than fifty years ago. The necessary high degree of structural perfection is achieved for a limited number of crystals, such as silicon, germanium, and related families; almost no observations of the dynamic effects have been made in metals. Studies of X-ray anomalous transmission in the transition metals are of considerable interest, particularly in tungsten which has a simple structure and a high absorption coefficient.

A method of electron-beam floating zone melting (EBFZM) is widely used to grow single crystals of high-purity refractory transition metals for years (Pfann, 1966; Shah, 1980). The growth of the perfect single crystals of the refractory metals presents difficulties because of the low defect formation energy and the stringent constraints on the level of the temperature gradients. The single crystals of molybdenum and tungsten, grown from the melt by this method, tend to have the specific substructure, characterized by the high dislocation density, reaching 10^5-10^7 cm^{-2}. The main part of these dislocations is collected in the walls, forming the dislocation substructure of the three orders of magnitude. On the one hand, the substructure is due to polygonization of dislocations arising during the growth by one of the known mechanisms (Bolling & Finestein, 1972, Kittel, 1996, Nes & Most, 1966). On the other hand, there is inevitable inheritance of the seed crystal substructure in the growing single crystal, which consists in the fact that the favorably oriented low-angle boundaries grow up into a crystal. Considerable efforts have therefore been made to improve the structural quality of the tungsten single crystals (Cortenraad *et al.*, 2001a). Modern methods of preparing the tungsten single crystals can produce the specimens having the dislocation density of about 10^5 cm^{-2}. A chemical composition, a growth rate, a number of passes by the liquid zone, geometry of the crystals and some other parameters of the growth in varying degrees affect the substructure, but in any case, the substructure of the crystals of molybdenum and tungsten grown from the melt is far imperfect (Glebovsky *et al.*, 1988; Glebovsky & Semenov, 1993-1994, 1995, 1999)

However, the single crystals, free of the specific substructure, can be grown by the secondary recrystallization process, which consists of the plastic deformation procedure and the high-temperature annealing procedure. The plastic deformation procedure of monocrystalline specimens can be produced by rolling in the vacuum rolling machines. The high temperature annealing procedure can be performed with the help of the anneal devices inside the rolling machines or in the EBFZM set-ups. The studies of structural perfection of the single crystals grown from the melt and by recrystallization are made by using the methods of X-ray rocking curves and angular scanning topography. To monitor the subgrain substructure of the tungsten single crystals, the X-ray anomalous transmission method has been employed as well. The optimal recrystallization process involves the deformation of single crystals with the [111] growth axis by rolling along the (112) plane. The 6-12% deformation is found to be optimal to get the polycrystals with the large grains of high perfection. The vacuum conditions are most suitable for vacuum rolling to avoid oxidation of the crystal surfaces during deformation at high temperatures. As a result, the single crystals of molybdenum and tungsten have the substructure which is characterized by both the record-low dislocation density and the small mosaic (Glebovsky & Semenov, 1999). For comparison, the tungsten single crystals, grown from the melt, contain the subgrains of the first order, elongated along the growth axis with the misorientation angles of 8-10′ of an arc. The crystallographically perfect tungsten single crystals, obtained by recrystallization, do not contain the subgrains of the first and second orders at all, and the maximum misorientation angles of the subgrains of the third order are less than 1′ of an arc. The structural changes in the perfect single crystals as a result of the thermal stresses, when they have been used as the seed crystals for growing the single crystals from the melt, have been studied.

It is well known that ideal growth techniques and technologies do not exist. All methods and technologies have their own advantages and disadvantages, so the main tasks of researchers consist in developing the advantages and in reducing negative effects of the disadvantages. The EBFZM method has its unique advantages which open wide prospects in the production of the high-purity refractory metals. It would be a mistake if the prospects will not be realized because of the complexities associated with the structural features of the single crystals grown from the melt.

The idea of the chapter is to show the most reliable ways of improving the structural quality of single crystals of the high-purity refractory metals. The recrystallization example for the tungsten single crystals shows how perspective and reliable are these ways in obtaining the structurally perfect single crystals.

2. Brief comments on the electron beam float-zone melting and growing single crystals

Zone-melting techniques (particularly, the EBFZM method) are useful for metals which are very reactive in the liquid state at high temperatures, so they cannot be processed at any contact with other materials (Pfann, 1966). Quoting Pfann, discovered the zone melting technique: "I regard the conception and development of zone melting as an exiting scientific advance. And I cannot help being saddened to hear it occasionally referred to as simply a technical innovation that was mysteriously evoked by the need for transistor grade germanium and silicon. I regard zone melting as elegant both in its simplicity and its surprising complexity." By far the EBFZM method, which is the crucibleless zone melting technique, is characterized by simplicity and complexity, but it is still the best one for melting refractory metals and their alloys. There are some well-known advantages of the method: small volume of the melt; the high efficient EB guns; the well-defined thermal gradients; no contamination from the crucible materials – because instead of them the surface tension of liquid metals works; the non-contaminating way of heating - the electron beams; effective purification which can be achieved due to evaporation of impurities in a vacuum. In parallel, there are some disadvantages of the EBFZM method: it can only be used in a vacuum; limitation of sizes of crystals due to surface overheating and thus decreasing the surface tension of the liquid zone; the high axial temperature gradients in a solid; the high thermal stresses and the high density of the unremovable dislocations; limitation of geometry and mass of the crystals produced by the EBFZM method. These or other properties of the method are marked as the merits or shortcomings, it would be wrong to perceive clearly. Thus, one of the major advantages of the method is absence of refractory crucibles and holding the liquid zone by the surface tension. However, high sensitivity of the surface tension to the surface-active impurities and the temperature gradients converts the recognized advantage into a serious drawback, which prevents growing the single crystals of large diameters, because mass of the liquid zone is too large so that surface tension forces are able to hold it. A similar comment can be done concerning the temperature distribution and the temperature gradients. Certainly, the well-defined thermal gradients are the advantage of the method, but their high values lead to formation of the specific substructure in single crystals, which creates great problems for both the physics research and industrial application. The electronic heating is the really controlled non-contaminating way of heating, but it may only be used in a vacuum, which is also a kind of contradiction when discussing the advantages and the disadvantages of the method.

3. Features of the single crystals growth

The single crystals of the high-purity refractory metals are widely used in modern material science and technology (Alonzo et al., 1995; Calverly et al., 1957; Glebovsky et al., 1998; Hay et al., 1968; Liu & Zee, 1996; Moest et al., 1998). This necessitates both studying purification processes and developing advanced techniques of growing single crystals of high-purity refractory metals with modern electron beam (EB) guns (M. Cole et al., 1968; Glebovsky et al., 1986). Crucibleless techniques with electronic heating are extremely important for melting, studying and preparing refractory metals because of their high chemical reactivity in the liquid state. Early zone refining theories such as progressive freezing, zone refining, zone crystal growing, nonideal separation and optimization are studied and discussed elsewhere (Pfann, 1966; Shah & Wills, 1975). Equipment and technologies such as types of the electron guns used, the drive mechanisms, heating and cooling, floating-zone melting, and stirring are also discussed in detail elsewhere (Shah, 1980). In the EBFZM method the liquid zone is held in place between two vertical collinear solid rods by its surface tension (Fig. 1). Single crystals of high-purity refractory metals can be grown exclusively by EBFZM because of their extremely high melting temperatures and chemical reactivity (Calverly *et al.*, 1957; Hay *et al.*, 1968; Alonzo *et al.*, 1995; Moest *et al.*, 1998; Glebovsky *et al.*, 1998). This necessitates both studying purification processes and developing advanced methods of growing single crystals metals using modern electron beam guns (M. Cole *et al.*, 1968; Glebovsky *et al.*, 1986). The main purpose in this field is to study the real structure of single crystals as a function of the technological parameters of the EBFZM method (Langer, 1980; Riedle *et al.*, 1994, 1996).

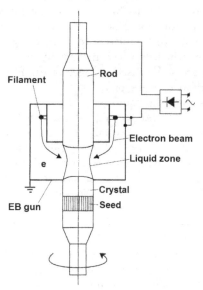

Fig. 1. Thermal zone of the EBFZM.

For effective melting and growing, the original EB guns have been elaborated on because the EB guns is the most important element of the EBFZM set-ups (Glebovsky *et al.*, 1986; Shah, 1980). In Fig. 2 the EB gun is shown, which consists of a cathode, an anode, and the focusing electrodes. The main features of the EB gun are: (a) the rod (crystal) serves as the anode, (2) the focusing electrodes made of a water-cooled copper, which makes the EB gun geometrically solid even at very high temperatures in the liquid zone, (3) the focusing electrodes form a stable circular electron beam field and focus it on the liquid zone, (4) the EB gun produces the well-defined thermal gradients on the crystal under the crystallization front. The cathode is made of a circular tungsten filament of 55 mm in dia. An arrangement of the focusing electrodes makes it possible to vary the electron-beam field from a diffuse pattern to sharp one. The advantage of the EB gun is its effectiveness at the refining and growing procedures during service of about 200 hours, compared to the known EB guns which can be used for no longer than 20-30 min. Thus, the original EB gun can be used continuously, both for refining of refractory metals and growing the single crystals at the growth rates of up to 50 mm/min, diameters up to 35 mm and lengths up to 1100 mm (Glebovsky et al., 1986). The growth of single crystals is usually accompanied by purifying liquid metals to high purity. It is demonstrated by preparation of the high-purity refractory metals with the residual impurities at the level of detection of the modern analytical techniques (Alonzo et al., 1995; Bdikin et al., 1999; Bozhko et al., 2008; Chaika et al., 2009; Brunner & Glebovsky, 2000a, 2000b; Cortenraad et al. 2001a, 2001b, 2001c, 2001d; Ermolov et al., 1999, 2002; Glebovsky et al., 1998; Markin et al., 2006, 2010; Moest et al., 1998; Shipilevsky & Glebovsky, 1989). The most problematic metals in growing the single crystals of the refractory metals (molybdenum, tungsten, niobium, and tantalum) are two - molybdenum and tungsten. Therefore, the focus of this chapter is devoted to just these two metals, although almost all the results can be easily applied to other two metals - niobium and tantalum.

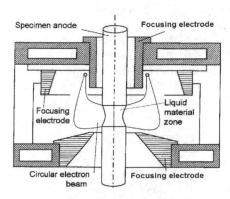

Fig. 2. Circular EB gun with focusing electrodes made of water-cooled copper.

4. Substructure of the molybdenum and tungsten single crystals

Single crystals of refractory metals with the relatively simple *bcc*-lattice grown by EBFZM have the specific dislocation substructure with the size of subgrains, which can be divided into three orders of magnitude. Table 1 shows approximate parameters of the substructures. The chemical composition, especially the content of the interstitial impurities, the growth rate, the number of the liquid zone passes, geometry of the crystal and other parameters significantly affect structural perfection of crystals.

There are several mechanisms of appearance of both the dislocations and substructure of single crystals at growing from the melt (Hurle, 1977; Reid, 1966): under influence of thermal stresses during growing and cooling of single crystals, due to the impurity concentration gradients in the solid phase, due to supersaturating of the lattice with vacancies, inheritance of the substructure of the seed crystal into the growing crystal. In reality, apparently, several mechanisms can operate simultaneously, or some of them will dominate. Obviously, in single crystals of the sufficiently pure refractory metals, effect of impurities on the substructure is unimportant (Akita et al., 1973). However, despite of a large number of studies in this area still remain unclearness related with influence of some factors in formation of the substructure.

Order of substructure	Average size of subgrains	Misorientation angles between subgrains
First order	1 mm < d < 8 mm	$30' < \theta < 4^0$
Second order	50 μm < d < 1 mm	$30'' < \theta < 30'$
Third order	0 < d < 50 μm	$0 < \theta < 30''$

Table 1. Estimated parameters of the crystalline substructure.

In growing the single crystals of molybdenum and tungsten by EBFZM one of the main monitored parameters is the growth rate (or the rate of liquid zone traveling, or the rate of the EB gun displacement). The growth rate is essential both at the crystallization stage, and at the post-crystallization annealing stage, which begins at an interface between the liquid zone and single crystal. At high temperatures, dislocations, regardless of nature of their origin, have very high mobility, so that there polygonization of the dislocation substructure has taken place. Apparently, along with increased dislocation mobility the stresses at the interface between the solid and liquid phases also contribute to formation of the polygonized structure. The typical substructure of the tungsten single crystal with the growth axis [001], revealed at the (010) plane parallel to the growth axis, is shown in Fig. 3. It is clearly seen that the boundaries of the subgrains of the first order extend for the considerable distances along the growth axis. They represent the walls or the dislocation network formed by potential for the *bcc*-lattice dislocations with the Burgers vectors b_1 = a/2 [111] and b_2 = a [100].

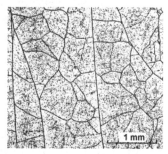

Fig. 3. Substructure of the tungsten single crystal with the vertical growth axis [001] grown at the rate of 2 mm/min. Electrolytic etching in 25% solution of NH_4OH.

The impurities and doping have significant effect on the substructure of single crystals grown from the melt. When microalloying occurs at the definite growth rate, the flat crystallization front is quite stable, and concentration supercooling does not develop. Changes in the dislocation substructure are associated with increase of the dislocation density inside subgrains, decrease of the average size of subgrains and increase of the misorientation angles between subgrains. In the case of doping in significant concentrations, the growth of single crystals becomes impossible at any growth rate. Figure 4 shows the longitudinal and transverse cross-sections of the polycrystalline Mo-2%W alloy, grown from the melt by EBFZM. This polycrystalline ingot contains the large grains elongated along the growth axis. Figure 5 shows the microstructure of the polycrystalline molybdenum ingot of low purity, also grown from the melt by EBFZM. The resulting structure is distinctive: the single-crystalline core surrounded with the polycrystalline periphery.

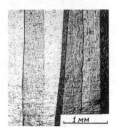

Fig. 4. Microstructure of the polycrystalline Mo-2%W alloy, grown from the melt by EBFZM (longitudinal and transverse cross-sections).

Fig. 5. Microstructure of the polycrystalline molybdenum ingot of low purity, grown from the melt.

To study effect of the growth rate on the substructure, the single crystals of molybdenum and tungsten, having the growth axes along the [001] axis, are grown. The growth rate varied from 0.2 to 50 mm/min, the rotation rate of the growing crystal - from 0 to 100 revolutions per minute. The range of the growth rates under study actually covers all growth rates, implemented in the EBFZM method: the lower limit depends on intense metal evaporation, and the upper limit – by possibility of full melting and stability of the liquid zone. The single crystals are grown in three passes of the liquid zone: the first pass at the growth rate of 6 mm/min, the second pass at the growth rate of 2 mm/min, the third pass - on the seed crystal at the rate from the above range of the rates.

4.1 The substructure of the molybdenum single crystals

The substructures of the molybdenum single crystals vary seriously depending on the growth rate (Glebovsky et al., 1988; Glebovsky & Semenov, 1994, 1995; Glover et al., 1970, Liu & Zee, 1996). At high growth rates the substructure is characterized by the high dislocation density and the more stressed state. The latter is confirmed by zone annealing the molybdenum single crystal of 8 mm in dia at the temperature close to the melting temperature and at the traveling rate of the EB-gun of 0.5 mm/min. At the low growth rates the dislocations have enough time to be polygonized, because single crystals stay at elevated temperatures for a longer time. Studies of the molybdenum single crystal grown at the growth rate of 6 mm/min show that after zone annealing the crystal has been polygonized to greater extent.

The single crystals grown at the growth rate of 0.5 mm/min have the specific developed substructure with an average size of subgrains of the second order of about 100 μm and the dislocation density, calculated from the etch pits, of 3×10^5 cm^{-2}. The substructure of the single crystal grown at 6 mm/min is characterized by the individual etch pits and the lack of the polygonized boundaries. The dislocation density is higher - of 1×10^6 cm^{-2}. The specific substructures of the molybdenum single crystals grown at different growth rates are shown at Fig. 6, 7, and 8. Most clearly the dependence of the misorientation angles of the subgrains on the growth rate is detected by the divergent X-ray beam patterns (Fig. 9). The misorientation angles of the subgrains of the second order in accordance with the value of discontinuities on the line (400) is 50′ of an arc for the growth rate of 0.5 mm/min, and 20′ of an arc – for the growth rate of 6 mm/min.

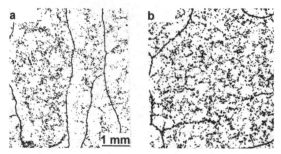

Fig. 6. Specific substructures of longitudinal (a) and transverse (b) cross-sections of the molybdenum single crystal, grown at 2 mm/min.

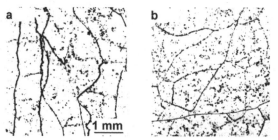

Fig. 7. Specific substructures of longitudinal (a) and transverse (b) cross-sections of the molybdenum single crystal, grown at 10 mm/min

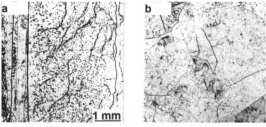

Fig. 8. Specific substructures of longitudinal (a) and transverse (b) cross-sections of the molybdenum single crystal, grown at 40 mm/min.

Fig. 9. Divergent X-ray beam patterns of the molybdenum single crystals, grown at rates: a- 0.5 mm/min, b- 6 mm/min.

Another feature of the substructure of the single crystals of molybdenum is radial heterogeneity, detectable at all growth rates. In the central part of the single crystals the boundaries are almost absent; however, at the periphery is observed intense polygonization. Radial inhomogeneity of the single crystals of 8 mm in dia is clearly visible on the divergent X-ray-beams patterns, where the value of shifts at edges of the line (310) is bigger than in the central part: at the edges – 50' of an arc, in the center – 20' of an arc. Dependence of nature of the substructure of the single crystals on radial inhomogeneity as well as on the growth rate can be explained by dislocation motion in the non-uniform temperature field of the single crystals.

4.2 The substructure of the tungsten single crystals

The typical substructures of the tungsten single crystals with the growth axis [100] and 10-12 mm in dia, grown at growth rates from 0.5 mm/min up to 40 mm/min, are shown in Fig. 10, 11, and 12. At the growth rates of 0.5-4 mm/min on the transverse cross-sections in the (001) plane, the distribution of the subgrains of the second order by size is close to normal. On the longitudinal cross-sections in the plane (100), pronounced elongated subgrains along the growth axis are found. At 40 mm/min, the polygonization process is not completed: the dislocation density inside the subgrains increases. At two different sites of the tungsten single crystal grown at the growth rates of 2 mm/min and 40 mm/min, the dislocation density increases by a half of an order of magnitude - from 8×10^4 cm^{-2} to 2×10^5 cm^{-2}. There is also decrease of the average size of the subgrains, and the misorientation angles are increased to $3\text{-}4^0$ of an arc. Apparently, such influence of the growth rate on the substructure of the tungsten single crystals is common for all refractory transition metals.

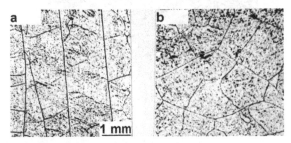

Fig. 10. Specific substructures of longitudinal (a) and transverse (b) cross-sections of the tungsten single crystal grown at 0.5 mm/min.

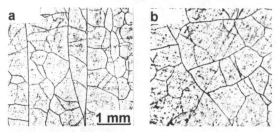

Fig. 11. Specific substructures of longitudinal (a) and transverse (b) cross-sections of the tungsten single crystal, grown at 2 mm/min.

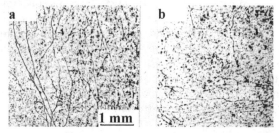

Fig. 12. Specific substructures of longitudinal (a) and transverse (b) cross-sections of the tungsten single crystal, grown at 40 mm/min.

Note significant irregularity of the substructure of the tungsten single crystals in the radial direction: the central part and the periphery are notably different, similar to differences observed in the molybdenum single crystals. The metallographic studies of the substructure of the single crystals of molybdenum and tungsten show that between them there is fundamental similarity. The rate of rotation, and the focusing and power fluctuations of the electron beam, leading to emergence of the constrictions and other defects on the surface of the single crystals, generally have no noticeable effect on the substructure of the single crystals. The fact that the temperature field is non-uniform in the radial direction has been confirmed by existence of the curvilinear crystallization front in vicinity of the seed during zone melting (Fig. 13). In turn, heterogeneity of the temperature field leads to non-uniform mechanical stresses which have different effects on the dislocation motion rate.

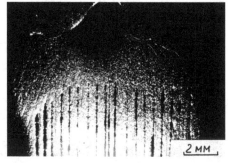

Fig. 13. Macrostructure of the longitudinal thin section of the tungsten single crystal of 11 mm in dia, showing the curvilinear crystallization front in vicinity of the seed.

Although unevenness of the substructure of single crystals creates some problems for physical studies and practical applications (Markin *et al.*, 2006, 2010; Mundy *et al.*, 1978), there are recent results on successful use of the as-grown single crystals for the manufacture of the STM tips (Bozhko *et al.*, 2008; Chaika *et al.*, 2009). It is demonstrated the main advantage of the single crystalline W[001] STM tips: sharpness, stability, and the predictable atomic structure. With these tips a set of the complimentary atomically resolved images of the complicated Si(557)5x5 stepped surface reconstruction is reproducibly received and revealed its atomic structure. The example of instability of the W[001] tip illustrates how the known tip axis orientation and the apex atom jump lengths may allow one to predict the atomic structure of the real single crystalline tip that can be of high importance for correct interpretation of the ultimately high resolution STM data. Nevertheless, presence of the specific substructures in single crystals, which prevents expansion of the practical application of single crystals, while insoluble problem for the researchers, that stimulates search for ways of obtaining the perfect single crystals of these metals.

5. Effect of thermal stresses on the substructure of the single crystals

The EBFZM method is characterized by presence of the high axial temperature gradients, especially near the solidification front in both the solid and liquid phases. Since the temperature gradient is a nonlinear function of the distance from the solidification front, *i.e.*, $d^2T/dz^2 \neq 0$, this leads to thermal stresses in growing single crystals that can cause multiplication of the dislocations. For a cylindrical crystal with dissipated radiating crystallization heat, the axial temperature gradients can be estimated from the known formulas relating the temperature, the black-body coefficients, the Stefan-Boltzmann constant, the crystal diameter, and thermal conductivity (Table 2).

Metal	The distance from the liquid zone Z, cm						
	0	0,2	0,4	0,6	0,8	1,0	1,2
Molybdenum	-618	-557	-505	-461	-422	-389	-359
Tungsten	-1453	-1207	-1021	-877	-763	-671	-596

Table 2. Axial temperature gradients (K/cm) in the solid state for single crystals of molybdenum and tungsten

The temperature along the axis of single crystals has been measured by optical micropyrometry. The holes of 1 mm in dia and 7-8 mm in depth, located along the axis of the single crystals, serve as a black-body model. Due to the high axial temperature gradients, the black-body model is appeared to be essentially non-isothermal. This increases the temperature measurement error up to $\pm 100^0$, but it is still possible to obtain the reproducible temperature profiles for all metals studied. Figure 14 shows the temperature distribution along the axis of the cylindrical tungsten single crystal of 15 mm in dia. It is seen that near the crystallization front the temperature along the crystal falls particularly sharply to 2000K at 40 mm from the front. The control of the temperature profile along the axis of the liquid zone is performed by measuring an electron current. Owing to the design of the EB gun, the temperature profile of the liquid zone can be effectively changed from diffuse one to sharp, which gives additional opportunity to manage the crystal growth.

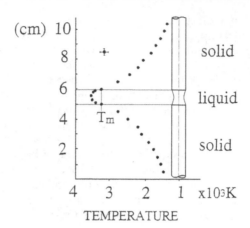

TEMPERATURE

Fig. 14. Temperature distribution along the axis of the cylindrical tungsten single crystal of 15 mm in dia.

For the case of the ax-symmetric temperature distribution along the cylindrical single crystal, resulting thermal stresses are mainly determined by the axial temperature gradients. At pre-melting temperatures, an elastic limit of metals is practically zero and thermal stresses are completely removed by the dislocations, *i.e.*, there is plastic deformation. Since at the growth of crystals from the melt the crystallization heat should released, then inevitably there is the temperature gradient in the solid phase, which leads to the certain density of unremovable dislocations (M. Cole, *et al.*, 1961; Buckley-Golder & Hurphreys, 1979; Esterling, 1980; Nes & Most, 1966; Otani, 1984;). The estimate of this dislocation density can be performed by:

$$\rho \geq \alpha \, gradT \, / \, b, \qquad (1)$$

Here, α – a linear coefficient of thermal expansion, K^{-1}; b - the Burgers vector, cm; *gradT* - the temperature gradient at the crystallization front, K/mm. For tungsten single crystals, such evaluation reveals that in order to have the dislocation density of $\rho = 10^4$ cm^{-2}, the temperature gradient must be less than 50 K/cm. Since the real temperature gradients are usually higher for an order of magnitude, in the melt-grown tungsten single crystals the dislocation density is typically 10^5-10^6 cm^{-2} or even more. The dislocation density in the boundaries is usually for an order of magnitude higher than in the bulk of the subgrains. The estimate by means of the formula (1) with the temperature gradient of 100 K/mm gives the dislocation density of $\rho \geq 5 \times 10^5$ cm^{-2} (Table 3).

Dislocation density ρ, cm^{-2}	Temperature gradients, K/см	
	Molybdenum	Tungsten
10^6	2900	3600
10^5	290	360
10^4	29	36

Table 3. Numerical estimates of temperature gradients at the crystallization front for single crystals of molybdenum and tungsten of 16 and 11 mm in dia, respectively.

Note that dislocations in single crystals are formed under action of thermal stresses in the growth process, and in the cooling process as well. Depending on the cooling rate, the number of the imposed dislocations in the growing crystal can be even higher than the number of the dislocations appeared during the growth (Nes & Most, 1966). The result is in decreasing subgrains sizes, as observed in the thermal shock at welding up both the seed crystal and initial rod together or cutting the crystal by the electron beam. In addition to the axial temperature gradients, quantitative estimates for the stationary stage of the crystal growth have shown, that to form the substructure, the cooling rate is important as well, which is realized in the growing and cooling processes when the crystal growth process is over. For example, in aluminum single crystals obtained by zone melting, the residual dislocation density is of $\rho \sim 10^2$ cm^{-2} when cooled to room temperature at $\sim 10^{-3}$ K/s (about a week), while at the increasing cooling rates on an order of magnitude, the residual dislocation density increases to $\rho \sim 10^4$ cm^{-2}.

In the case of growing the single crystals of molybdenum and tungsten by EBFZM, one can get the estimate from below for the maximum cooling rate of the crystal growth process. The maximum of this magnitude occurs in the solid phase just below the crystallization front. Multiplying the axial temperature gradient on the growth rate, one obtains the value of the maximum cooling rate of the crystal during the growth:

$$\frac{dt}{dz} \times \frac{dz}{dt} = \frac{dT}{dt}. \tag{2}$$

Taking the data of the temperature gradients (Table 3) and the growth rate of 2 mm/min, the most frequently used in practice, one can obtain the cooling rate for molybdenum and tungsten, respectively, 2K/s and 5 K/s, i.e., the cooling rates are very high, taking into account the corresponding values for aluminum (10^{-3} K/s). It should be noted that significant reducing the crystallization rate in the EBFZM method is impossible. The fact that molybdenum and tungsten have very high vapor pressure at $T > T_m$ and at the growth rate of 0.5 mm/min, the metal losses by evaporation can reach 30% of initial mass.

However, even if the crystallization rate tends to zero, desired reduction of the dislocation density still can not be achieved. The reason is that the crystallization rate never coincides with the growth rate (or the rate of the EB gun displacement). In fact, the crystallization rate affects significantly the hydrodynamic processes developing in the liquid zone (Kobayashi, 1970; Kobayashi & Wilcox, 1982; Murphy, 1987; Surek & Chalmers, 1975). These processes give rise to oscillations in the growth rate; moreover, the instantaneous crystallization rate in these moments of time can be significantly greater than the cooling rate, as shown by estimates for molybdenum and tungsten. Presence of such oscillations of both the temperature and growth rate is shown elsewhere (Mullins & Sekerka, 1964; Wilcox & Fuller, 1965). The frequency of these oscillations is close to the inverse of the thermal time constant of the melt-crystal system:

$$f \approx \frac{a}{S} \tag{3}$$

Here, f – a frequency of oscillation, Hz; a – a thermal diffusivity, cm^2/s; S – a cross-sectional area of the crystal, cm^2. Oscillations of the crystallization front in presence of impurities in a crystal lead to substantial change in the distribution coefficient and, consequently, the so-called transverse striations in the crystals observed by autoradiography.

Considerable interest represents an estimate of the cooling rate from T_m to room temperature. The simplest case can be considered, when the one-dimensional quasilinear heat conductivity equation with the constant coefficients is numerically solved. The process of heat propagation in a homogeneous rod can generally be described by the equation:

$$\rho'C_p \frac{\partial T}{\partial \tau} = \frac{\partial \left(K \frac{\partial T}{\partial X} \right)}{\partial x} + f(x,t) \tag{4}$$

where $T(x, t)$ – a temperature at the point X of the rod at the moment of time t; C_P – heat capacity per unit mass at constant pressure; ρ_m – metal density; K – thermal conductivity; f – density of the heat sources (sinks); X – a coordinate along the rod length L. If one assumes that K, C_P, and ρ' are constant, the equation (4) can be rewritten as:

$$\frac{\partial T}{\partial t} = a^2 \frac{\partial^2 T}{\partial X^2} + f(X,t) \tag{5}$$

Thus, it is necessary to find the continuous solution at $T = T(x, t)$ of the equation (5) for

$$\overline{D} = \{0 \le X \le L; 0 \le t \le t^1\}$$

if

$$T(X,0) = T_0(X); 0 \le X \le L$$
$$T(0,t) = T_1(t); 0 \le t \le t_1$$
$$T(L,t) = T_2(t); 0 \le t \le t_2$$

For a uniform rod with the diameter d, cooling due to radiation and thermal conductivity after switching off the electron beam one obtains instead of the equation (5):

$$\frac{\partial T}{\partial t} = a^2 \frac{\partial^2 T}{\partial X^2} - \frac{4\varepsilon T^4}{d\rho'C_p} \tag{6}$$

In the task the most interesting is the value of $\partial T/\partial t$ with the limitations $0 \le X \le L$; and $t=0$. This value is calculated numerically, and it provides the cooling rate at the upper limit. For these calculations, the following numerical values of the physical parameters are used. For molybdenum: $K = 0.909$ Wcm^{-1}K^{-1}, $C_P = 0.235$ Jg^{-1}K^{-1}, $\rho_{Mo} = 10.2$ gcm^{-3}, $d = 1.6$ cm. For tungsten: $K = 0.945$ Wcm^{-1}K^{-1}, $C_P = 0.172$ Jg^{-1}K^{-1}, $\rho_W = 19.2$ gcm^{-3}, $d = 1.1$ cm. The cooling rate for molybdenum is 2×10^4 K/s, and for tungsten – 5×10^4 K/s. Although these values exceed the cooling rates at the stationary stage of the crystal growth, they can not significantly impact on deterioration of the substructure, because in 5-10 seconds for the most part of the crystal they become comparable with the cooling rates at the stationary phase. The dislocation density can increase only a few in the surface layer near the end of the crystal. Therefore, when the crystals of refractory metals are grown from the melt, it is absolutely impractical to cool slowly, from scientific or technological points of view.

6. Recrystallization of single crystals

Because of relatively low perfectness of single crystals of some semiconductors and refractory metals, grown by EBFZM, there are some studies made to improve the substructure of single crystals. It is well known that the growth of the semiconductor alloys crystals is one of old problems in physics and practice of the crystal growth. The studies on growing such crystals include the casting-recrystallizing-annealing procedures and require careful balancing of the pseudo binary melt stoichiometry, which inherently is a quite difficult process. Increase in recrystallization efficiency can be achieved by adjusting the casting conditions and the suitable thermal gradient during recrystallization (Yadava *et al.*, 1985). This means, that the mechanism of the crystal growth is a combination of both the chemical potential gradient and temperature gradient of zone melting processes, so the growing processes of semiconductor alloys crystals are complicated at their practical realization.

In comparison with the growth of semiconductor crystals, the growth of such simple metals like molybdenum and tungsten seems to be a relatively non-problematic task. However, the main obstacles are both the high melting temperatures and high temperature gradients along single crystals when growing from the melt. Because of these obstacles the growth of single crystals of refractory metals becomes the very complicated task which attracts attention of many scientists for a half a century. Numerous studies concerning this problem have shown that these metals, in spite of their crystallographic simplicity, require special studies, knowledge, and equipment.

Using collected experimental information, the single crystals of molybdenum and tungsten, free of the specific substructure, are produced by recrystallization including plastic deformation and high-temperature annealing (Bdikin *et al.*, 1999; Katoh *et al.*, 1991). Plastic deformation of the crystals has been done by rolling in the vacuum rolling machine or in the standard mills in air. High temperature annealing is produced with the help of the heating devices located inside the rolling machine or in the set-up for electron-beam zone melting and growing single crystals. The comparison of structural perfection of the single crystals grown from the melt and grown by recrystallization is done by both X-ray rocking curves and angular scanning topography.

As a rule, the recrystallized single crystals of molybdenum and tungsten have the substructures characterized by both the record-low dislocation density and small-angle mosaic. It should be noted that the lower dislocation density in the single crystals can only be achieved by recrystallization. As mentioned already before, the optimal procedure involves deformation of the single crystals with the <111> growth axis by rolling along the (112) plane (Bozhko *et al.*, 2008). To monitor both the substructure and perfection of the tungsten single crystals, the anomalous X-rays transmission has been employed as well.

6.1 Experiments on recrystallization of the tungsten single crystals

The tungsten single crystals under study are grown by the EBFZM method (Glebovsky *et al.*, 1986). The single crystals have the different crystallographic growth axes. The high-purity tungsten powders of chemical purity 99.99% are used as a starting material. The as-grown single crystals are 11-22 mm in dia and 100 mm in length. The structural studies involve the X-ray diffraction microscopy methods, namely, angular scanning topography and rocking curves (Aristov *et al.*, 1974; Bozhko *et al.*, 2008; Brunner & Glebovsky, 2000a, 2000b; Ermolov *et al.*, 1999, 2002; Riedle *et al.*, 1994, 1996).

Several deformation systems are studied, which differ by the crystallographic parameters and strain, because two parameters such as the growth axis and the deformation direction are most important to get the strained single crystals before high-temperature annealing. Crystallographic systems tested are [100]/(010), [100]/(011), [110]/(110), [110]/(111)], [111]/(110) and [111]/(112) where [growth axis]/(rolling plane). The 6-12% deformation by rolling of the two-cant bars with fixed crystallography is found to be optimal to get the large grains. The vacuum conditions are most suitable for the crystal deformation by rolling because they enable one to avoid oxidation of the crystal surfaces during deformation at high temperatures.

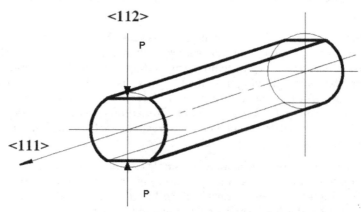

Fig. 15. Two-cants bar with fixed crystallography for rolling. The crystallographic parameters are for the system [111]/(112) where [111] is the growth axis and (112) is the rolling plane (P).

Sample	Growth axis	Rolling plane	Strain, %	Size of grains
1	[100]	(010)	13.4	Grains, 8 mm
2	[100]	(011)	11.6	Grains, 8 mm
3	[100]	(010)	12.3	Grains, 8 mm
4	[100]	(011)	11.6	Grains, 8 mm
5	[110]	(110)	7.9	Grains, 15 mm
6	[110]	(111)	7.5	Grains, 15 mm
7	[111]	(110)	7.5	Grains, 15 mm
8	[111]	(112)	6.6	Grains, 25 mm

Table 4. Parameters of the samples subjected to vacuum rolling followed by high-temperature annealing.

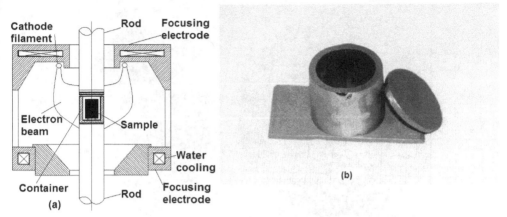

Fig. 16. Electron-beam annealing of a single crystal in a container, a-EB gun, b-container.

The billets to be plastically deformed are produced from the cylindrical tungsten single crystals and have a form of the two-cant bar so that crystallography has been fixed before rolling (Fig. 15). Plastic deformation is performed in one pass at the temperature 900⁰C in the vacuum rolling machine or, in a number of cases, in the standard rolling mill in air. In Table 4 are listed the main data for the recrystallized single crystals. The hypercritically strained tungsten single crystals are annealed at 2500⁰C in the EBFZM set-up by the defocused electron beam (Fig. 16, a). In order to obtain the more uniform temperature field in the specimens, the tungsten container is used. Its height is 40 mm, the diameter 30 mm, the wall thickness 3 mm (Fig. 16, b). The tungsten specimen of the maximal length 30 mm and the diameter 20 mm is installed inside the container on the tungsten holders. The container has been arranged coaxially with the EB gun equipped with a circular cathode.

6.2 Recrystallized tungsten single crystals in comparison with ones grown from the melt

In order to elucidate influence of recrystallization on the real substructure of tungsten single crystals, the topograms of angular scanning of the specimens, cut from the melt-grown single crystal before deformation, are taken. In Fig. 17, the topograms show the mosaic substructures of two as-grown tungsten single crystals grown in identical conditions. The size of the subgrains is about 1-2 mm, the misorientation angles between subgrains are about 50″ of an arc. On the angular scanning topograms are well seen the small-angle boundaries, their misorientation angles make up tens of the angular minutes, for some single crystals they can exceed 1° of an arc. Single crystals of such structural quality are not suitable for producing, i.e., deflectors and targets using in the runs concerned with channeling of high-energy particles beams. It is necessary to decrease substantially the dislocation density in such crystals and to deplete them from the small-angle boundaries. In contrast to the melt-grown single crystals, necessary structural perfection can only be achieved by the recrystallization processes consisting of plastic deformation and high-temperature annealing.

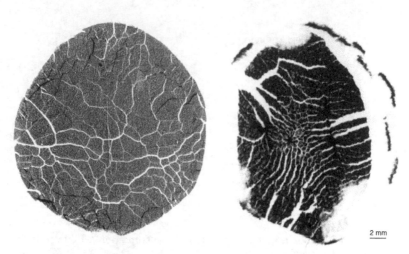

2 mm

Fig. 17. Angular scanning topograms of two as-grown tungsten single crystals, (110) reflection, CuK_α.

The rocking curves are recorded in the dispersion-free system using the perfect silicon single crystal as a monochromator (Kittel, 1996). To measure the rocking curves, the angle between the X-ray source and the detector is fixed under the first order Bragg diffraction angle. A crystal is rotated with respect to the incoming X-ray beam, and an intensity of the diffracted beam is measured as a function of the angle. The X-ray diffraction rocking curves are measured in three different sites on a specimen surface. The sites differ in a rotation angle of 45^0 in order to exclude texture effects. A half-width (FWHM) of the rocking curves is ~1^0 of an arc. Different subgrains show up as individual peaks on the rocking curves. The angular scanned topograms are obtained in the Θ-2Θ scanning regime and used to study the position, dimensions, and misorientation angles of subgrains. This X-ray diffraction technique is based on a principle of the Bragg reflection (Bdikin et al., 1999; Cortenraad et al., 2001a, 2001b, 2001c, 2001d).

In Table 1 are listed main data for those single crystals. The final recrystallized specimen is the polycrystal incorporating the distinct large grains (Fig. 18). After the recrystallization procedure the large grains do not contain the small-angle boundaries and have the relatively low dislocation density. The optimal procedure, involving deformation of the single crystal with the <111> growth axis by rolling along the (112) plane, is the most suitable for recrystallization. The ~6% deformation is found to be optimal for these crystallographic parameters to get few single nuclei and then large grains to be grown. Higher deformation leads to too many nuclei and as a result - to smaller grains grown. The vacuum conditions are most suitable for rolling because they enable one to avoid oxidation of the crystal surfaces during deformation at high temperatures. By this technique the high purity single crystals of the low dislocation density and free of the small angle boundaries are produced. In several cases the rolled specimens are annealed outside the container. That results in formation of the surface damaged layer of up to 500 µm thick. The middle part of such specimens is virtually free from the small-angle boundaries and has the perfect structure although there are many small subgrains at the periphery.

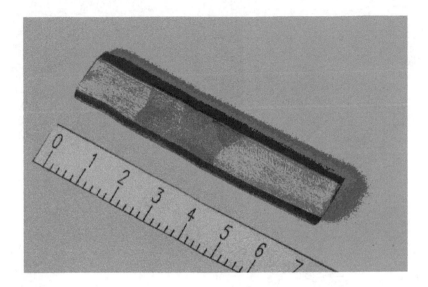

Fig. 18. Recrystallized tungsten polycrystal incorporating three distinct large grains.

Using a tungsten container for high-temperature annealing enables both avoiding damage of the specimen surface layer and decrease of the temperature gradient. The specimens with record structural perfection are obtained by using this technique. The angular scanning topogram and corresponding rocking curve, taken from the specimen annealed in the container are shown in Fig. 19. The angular scanning topogram shows the perfect surface of the specimen, without any boundary or other defects (compare with the topogram for as-grown specimens in Fig. 17). The rocking curve has only one sharp peak with the width at a half height (FWHM) of ~50" of an arc. Correspondingly, the dislocation density is about $5x10^4$ cm^{-2} for this specimen.

Strong changes of the substructure of the perfect tungsten single crystal has been found when the latter has been used as a seed crystal for growing a new tungsten crystal from the melt by EBFZM (Glebovsky & Semenov, 1999). Before growing, the dislocation density of the seed crystal is of about 10^4 cm^{-1}, and it does not contain any subgrains. After growing from the melt, the substructure of the seed single crystal indicates significant deterioration - subgrains of 500 μm are appeared, and they are elongated along the growth axis with the misorientation angle of 8-10' of an arc (Fig. 20). At the growth of single crystals of refractory metals from the melt, the main contribution to formation of the substructure is made by the dislocations arising under influence of the thermal stresses during the growth and cooling of the growing crystal. During growing from the melt, dislocations may arise due to thermal stresses in the solid and under action of the impurity concentration gradients due to lattice oversaturation with vacancies. Subgrains are formed by fresh dislocations and walls in spite of the fact that the seed crystal does not contain any subgrains before growing.

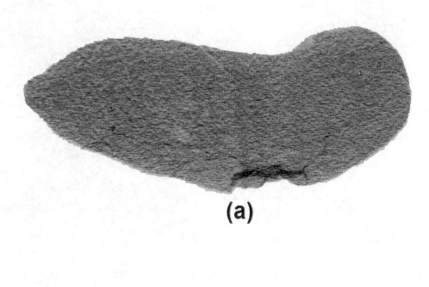

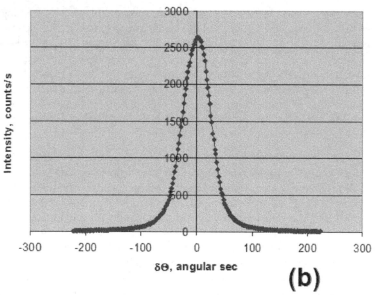

Fig. 19. Angular scanning topogram (a) and rocking curve (b) of the tungsten single crystal, annealed in the container; the surface plane (110).

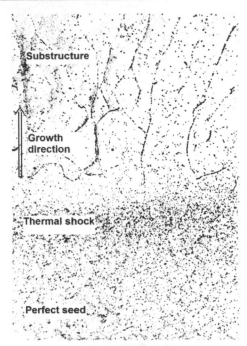

Fig. 20. Appearance of the fresh dislocation walls in the tungsten single crystal during the growth from the melt.

6.3 Anomalous X-ray transmission in the tungsten single crystals

A diverging X-ray beam is the beam where the angles of incidence of the X-rays on the specimen differ for different parts of the specimen (Bdikin *et al.*, 1999). Thus, for the given wavelength the diffraction conditions are satisfied along a certain line on the surface of the specimen. The classical system is used to record anomalous X-ray transmission (Borrmann effect). A fine-focus tube with a focus measuring 50x50 μm^2 is used as a source of the diverging X-ray beam. The distance between the specimen and the photographic film is 10 cm and that between the source and the specimen is 18 cm (Fig. 21). An asymmetric extraction geometry (the distance between the source and the specimen is not equal to that between the specimen and the photographic film) is used to eliminate focusing of the diffracted rays over the radiation spectrum. In the case of dynamic diffraction (perfect crystal) at points, where the diffraction conditions are satisfied, amplification of intensity is observed in the transmitted beam. In kinematic approximation (crystal containing defects), intensity is suppressed. The fact that transmitted radiation is recorded, and radiation intensity is the same in the diffracted and forward transmitted beams, together with observation of amplification of characteristic $CuK_{\alpha1,2}$ radiation intensity in these directions indicate that diffraction is of dynamic nature. Observation of splitting of these beams into a $CuK_{\alpha1,2}$ doublet (Fig.22) and a shift as the specimen rotates indicate that these beams are of diffraction origin and are not the topological characteristic of the specimen. Diffractometering of the waves which passed in the directions R and T also confirms that intensities of the diffracted and transmitted beams are the same.

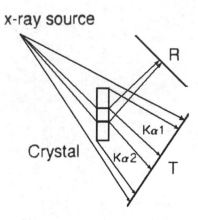

Fig. 21. Scheme for recording the anomalous transmission of X-rays. R-reflection, T-transmission.

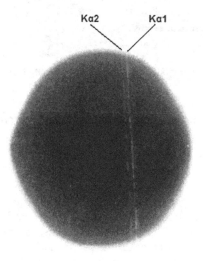

Fig. 22. Anomalous X-rays transmission in a perfect W single crystal; lines are from transmitted beams.

Calculations using the kinematic theory show that for Cu-radiation and the specimen thickness of 0.3 mm ($\mu t \sim 98$) absorption is 10^{43}. With the available radiation sources (no more than 10^7-10^8 pulse/s) it is impossible to obtain recordable transmission intensity. For Mo-radiation, absorption is lower than that for Cu-radiation. The linear absorption coefficient of the 0.3 mm thick tungsten single crystal is calculated to be $\mu t = 53.4$ on the Mo-radiation characteristics.

If the absorption coefficient is conceived as $\sigma = \mu_i t + y_i$, where μ_i is a coefficient of interference absorption and y_i is responsible for renormalization of the atomic amplitude of scattering, then for the perfect tungsten single crystal can be obtained $\mu_i = 369.2$, $y_i = -6.63$. The measurements of the absorption coefficient of a diffracted Mo-radiation yield $\sigma = 7.3$. Under assumption that y_i is independent of the defects concentration one obtains $\mu_i = 464.3$ cm^{-1}, i.e., in the real specimen the interference absorption coefficient is by 25% greater than it has to be in the perfect crystal.

The effects associated with dynamic X-ray propagation can only be observed in crystals for which the distance between the dislocations does not exceed an extinction length L. The calculations for (110) CuK_α reflection for the defect-free tungsten crystal give $L = 1.7$ μm, i.e., the critical dislocation density is $N_d = L^{-2} = 3.5 \times 10^7$ cm^{-2}. From the view point of the specific features of X-ray scattering, the crystal lattice defects fall into two classes. The first class involves localized defects, e.g. vacancies, which, virtually, do not deform the reflective atomic planes and, therefore, do not change the rocking curve width. Occurrence of such defects in the crystal only results in decrease of intensity of transmitted radiation, i.e., the growth of the absorption coefficient. The second class involves, for example, dislocations, presence of which gives rise to distortion of the atomic planes and violation of the crystal lattice period. These defects lead to increase of both the angular divergences of the diffracted beam (rocking curve width) and interference absorption coefficient. In the defect-free crystal the waves propagating in the crystal have the finite natural angular width $\delta = \lambda / L$, where λ is the characteristic radiation wavelength.

In order to determine the type of the defects which predominate in the perfect tungsten single crystals and their density, the rocking curves for the (110) reflection in the Bragg and Laue geometries are recorded, and measured the interference damping factor μ_i. In the Laue geometry the width of the rocking curve is 32″ of an arc whereas in the Bragg geometry the width of the rocking curve is 72″ of an arc. The calculated values for these widths in the defect-free crystal are 5.5″ of an arc and 48.6″ of an arc in the Laue and Bragg geometries, respectively. Hence, broadening of the rocking curves in the real crystal compared with the defect-free crystal is around 25″ of an arc for both rocking curves. The difference between the measured values of the rocking curve width and the interference absorption coefficient, and the calculated values for the defect-free crystal can be used to determine the dislocation density N_d in the crystals under study. Estimates of the dislocation density using the width of the rocking curve give $N_d = 2 \times 10^5$ cm^{-2}, and using the interference absorption coefficient – $N_d = 4 \times 10^5$ cm^{-2}. These data show a good agreement with the dislocation density determined using etch pits, $N_d = 5 \times 10^5$ cm^{-2}. Thus, the dislocation density in the real crystal is substantially lower than the critical density. Note that the dislocation density determined from the interference absorption coefficient is higher than that obtained from the rocking curve width. This is evidently attributable to presence of the defects in the crystal (such as the point defects) which increase absorption but make no contribution to the rocking curve width.

The technique used to study the Borrmann effect allows to obtain the angular scanned transmission and reflection topograms for the tungsten crystals, which reveal subgrains larger than 1-2 mm. Comparison of the transmission and reflection topograms suggests that the subgrain substructure exhibits similar reflection but the transmission topograms have a higher image contrast as a result of dynamic narrowing of the diffracted beam for the thick crystal (the smaller width of the rocking curve in the Laue geometry). The dislocation density determined by the interference absorption coefficient is larger than that determined by the rocking curve width. This is likely due to occurrence of the localized defects in the crystal, for example, the vacancies which virtually do not deform the reflective atomic planes. It is worth mentioning that the reflection topograms include the weak misorientations at the edges which on the transmittance topograms are not manifested since defectness is higher at the crystal edges than in the center. That excludes the effect of anomalous transmission.

The image of the mosaic subgrains is absent on the transmission topograms. This is because the concentration of the defects in the subgrains exceeds the critical concentration $N_d > L^{-2}$, and the regime of dynamic diffraction is not realized in them. Note that in the diffraction direction in the Bragg geometry, the subgrain formation and other structural features, leading to change in the direction of the diffracted beam, are clearly seen on images. This can be attributed to the fact that the distances between the dislocations in such boundaries are less than L and, accordingly, the small-angle boundaries are not transparent from a standpoint of dynamic diffraction that gives rise to the shadowing effect on the topogram.

The obtained results suggest the conclusion about the character of influence of recrystallization on the real structure of the tungsten single crystals. Importantly that effect of anomalous X-ray transmission manifests itself in the subgrains of ~1 mm in size in the as-grown single crystals as well. After recrystallization the sizes of the subgrains having the perfect structure appears to be much larger. In accordance with the dynamic theory of X-ray scattering, the calculated value of FWHM of the rocking curve for the perfect tungsten single crystal is 48" of an arc. Unfortunately, it is quite difficult to estimate the dislocation density in a limit of small broadening of the rocking curve. Thus, it is possible only to declare - the dislocation density in the tungsten single crystals is very low, about 10^4 cm^{-2}.

7. Conclusions

The dislocation substructure of single crystals of molybdenum and tungsten is characterized by significant similarity and remains virtually unchanged at the growth rates of 0.5-5 mm/min. Significant changes in the substructure, reflected in increasing fragmentation of subgrains and the misorientation angles of up to 3-4^0 of an arc, take place when the growth rates increase to 10-20 mm/min and above. After sudden increase of the growth rates from 2 mm/min to 40 mm/min, the dislocation density inside subgrains increases on an order of magnitude, reaching 5×10^6 cm^{-2}.

Due to the high temperature gradients near the crystallization front and the high cooling rates, the growth of the perfect single crystals of molybdenum and tungsten from the melt is impossible. Even using the perfect seed crystal, free from the small-angle boundaries with the misorientation angles 3' of an arc, the method does not allow growing crystals of satisfactory structural quality.

The numerical estimates show that the cooling rates of single crystals can reach 10^4 K/s. This results in small increase in the dislocation density in the thin surface layer during cooling, which is quite acceptable. Therefore, to cool single crystals slowly after growing by EBFZM is impractical from all points of view.

The conditions of the growth of single crystals with the ultimately low dislocation density and the small subgrains spread are revealed. The relatively low dislocation density and the lack of the small-angle boundaries are achieved by recrystallization. The optimal procedure involves the 6% deformation of the single crystalline specimen with the <111> growth axis by rolling in vacuum along the <112> plane and *in-situ* high-temperature annealing.

To monitor the subgrain substructure of the tungsten single crystals, the anomalous X-ray transmission method is effective. The Borrmann effect is observed in the recrystallized perfect tungsten single crystals. The dislocation density determined by diffraction data is close to that determined by etch pits (~5×10^{-4} cm^{-2}). This opens potentialities for controlling the dislocation density by the X-ray diffraction techniques. The perfect single crystals can be employed as the novel crystalline deflectors to monitor the beams of relativistic charged particles and for other applications. The recrystallization example shows how perspective and reliable is this way in obtaining structurally perfect single crystals of tungsten and other refractory metals as well.

8. Acknowledgement

The author has a great pleasure to express sincere acknowledgment to my colleagues and friends Valery Semenov, Sergey Ermolov, Eugene Stinov, Sergey Markin, Boris Shipilevsky, and Sergey Bozhko from the Institute of Solid State Physics, Chernogolovka, Russia, for favorable attitude, cooperation in science and life. The author is very grateful to Wolfgang Gust from the Max-Planck Institute fuer Metallforshung, Stuttgart, Germany, to Hidde Brongersma from the Technical University of Eindhoven, Eindhoven, The Netherlands, and to Wayne King from the Lawrence Livermore National Laboratories, Livermore, USA, for fruitful discussions and friendliness for many years. The skillful technical assistance of Victor Lomeyko from the Institute of Solid State Physics, Chernogolovka, Russia, is greatly acknowledged.

9. References

Akita, H., Sampare, D.S., & Flore, N.F. (1973). Substructure control by solidification control in copper crystals. *Metallurgical Transactions*, Vol. 4, pp.1593-1597, ISSN 1073-5623.

Alonzo, V., Berthier, F., Glebovsky, V.G., Priester, L., & Semenov, V.N. (1995). Study of microstructure of a Ni(100) single crystal grown by electron beam floating zone melting. *Journal of Crystal Growth*, Vol.156, pp.480-486, ISSN 0022-0248.

Aristov, V.V., Shmytko, I.M., & Shulakov, E.V. (1974). Application of the X-ray divergent-beam technique for the determination of the angles between crystal blocks. *Journal of Applied Crystallography*, Vol.7, No.4, pp.409-413, ISSN 1600-5767.

Bdikin, I.K., Bozhko, S.I., Semenov, V.N., Smirnova, I.A., Glebovsky, V.G., Ermolov, S.N., & Shekhtman, V.S. (1999). Observation of anomalous transmission of X-rays in tungsten single crystals. *Technical Physics Letters*, Vol.25, No.12, (December 1999), pp.933-935, ISSN 1063-7850.

Bolling, C.F., & Finestein, D. (1972). On vacancy condensation and the origin of dislocations in growth from the melt. *Philosophical Magazine*, Vol.25, No.45, pp.45-66, ISSN 1478-6435.

Bozhko, S.I., Glebovsky, V.G., Semenov, V.N., & Smirnova, I.A. (2008). Study on the growth of tungsten single crystals of high structural quality. *Journal of Crystal Growth*, Vol.311, No.1, pp.1-6, ISSN 0022-0248.

Brunner, D., & Glebovsky, V.G. (2000). The plastic properties of high-purity W single crystals. *Materials Letters*, Vol.42, (February 2000), pp. 290-296, ISSN 0167-577X.

Brunner, D., & Glebovsky, V.G. (2000). Analysis of flow-stress measurements of high-purity tungsten single crystals. *Materials Letters*, Vol.44, (June 2000), pp. 144-152, ISSN 0167-577X.

Buckley-Golder, J.M., & Hurphreys, C.J. (1979). Theoretical investigation of temperature distribution during Chochralski crystal growth. *Philosophical Magazine*, Vol.39, No.1, pp.41-57, ISSN 1478-6435.

Calverly, A., Davies, M. & Lever, R. (1957). The floating-zone melting of refractory metals by electron bombardment. *Journal of Scientific Instruments*, Vol.34, No.4, pp.142-144, ISSN 0950-7671.

Chaika, A.N., Semenov, V.N. Glebovsky, V.G., & Bozhko, S.I. (2009). Scanning tunneling microscopy with single crystalline W[001] tips: High resolution studies of Si(557)5x5 surface. *Applied Physics Letters*, Vol.95, No.17, 173107, ISSN 0003-6951.

Cole, M., Fisher, D.S., & Bucklow, J.A. (1961). Improved electron beam device for zone melting. *British Journal of Applied Physics*, Vol.12, No.10, pp.577-578, ISSN 0508-3443.

Cole, G.S. (1971). Inhomogeneities and their control via solidification. *Metallurgical Transactions*, Vol. 2, pp.357-370, ISSN 1073-5623.

Cortenraad, R.; Ermolov, S.N.; Semenov, V.N.; Denier van der Gon, A.W.; Glebovsky, V.G.; Bozhko, S.I., & Brongersma, H.H. (2001). Growth, characterization and surface cleaning procedures for high-purity tungsten single crystals. *Journal of Crystal Growth*, Vol.222, pp. 154–162, ISSN 0022-0248

Cortenraad, R., Ermolov, S.N., Moest, B., Denier van der Gon, A.W., Glebovsky, V.G. & Brongersma, H.H. (2001). Crystal face dependence of low energy ion scattering signals. *Nuclear Instruments & Methods in Physical Research, B*, Vol.174, pp.173-180, ISSN 0168-9002.

Cortenraad, R., Ermolov, S.N., Denier van der Gon, A.W., Glebovsky, V.G., Brongersma, H.H., Manenschijn, A., Gartner, G., & Belozerov E.V. (2001). Cleaning procedures for single crystal tungsten substrates. *Inorganic Materials*, Vol.37, No.7, pp.673-677, ISSN 0020-1685.

Cortenraad, R., Ermolov, S.N., Semenov, V.N., Denier van der Gon, A.W., Glebovsky, V.G., Bozhko, S.I., Stinov, E.D. & Brongersma, H.H. (2001). Electron-beam growing and purification of W crystals. *Vacuum*, Vol.62, pp.181-188, ISSN 0042-207X.

Ermolov, S.N., Cortenraad, R., Semenov, V.N., Denier van der Gon, A.W., Boghko, S.I., Brongersma, H.H., & Glebovsky, V.G. (1999). Growth and characterization of monocrystalline tungsten substrates. *Vacuum*, Vol.53, pp.83-86, ISSN 0042-207X.

Ermolov, S.N., Glebovsky, V.G., Cortenraad, R., Moest, B., Stinov, E.D., Denier van der Gon, A.W., & Brongersma, H.H. (2002). Low-energy ion scattering by various crystallographic planes of tungsten single crystals. *Physics of Metals & Metallography*, Vol.93, pp.443–449, ISSN 0031-918X.

Esterling, D.M., (1980) Dislocation dissociation in some *bcc*-metals. *Acta Metallurgica*, V.28, pp.1287-1294, ISSN 0956-7151.

Glebovsky, V.G., Lomeyko, V.V., & Semenov, V.N. (1986). Set-up for electron-beam zone melting of refractory materials. *Journal of Less-Common Metals*, Vol.117, pp.385-389, ISSN 0022-5088.

Glebovsky, V.G., Semenov, V.N., & Lomeyko, V.V. (1988) Influence of the crystallization conditions on the structural perfection of molybdenum and tungsten. *Journal of Crystal Growth*, Vol.87, No.1, pp.142-150, ISSN 0022-0248.

Glebovsky, V.G., & Semenov, V.N. (1993-1994), Electron-beam floating zone melting of refractory metals and alloys: art and science. *International Journal of Refractory Metals & Hard Materials*, Vol.12, pp.295-301, ISSN 0263-4368.

Glebovsky, V.G., & Semenov, V.N. (1995). Growing single crystals of high-purity refractory metals by electron-beam zone melting. *High-Temperature Materials & Processes*, Vol.14, pp.121-130, ISSN 0334-6455.

Glebovsky, V.G., Sidorov, N.S., Stinov, E.D., & Gnesin, B.A. (1998). Electron-beam floating zone growing of high-purity cobalt crystals. *Materials Letters*, Vol. 36, August, pp.308-314, ISSN 0167-577X.

Glebovsky, V.G., & Semenov, V.N. (1999). The perfection of tungsten single crystals grown from the melt and solid state. *Vacuum*, Vol.53, pp.71–74, ISSN 0042-207X.

Glover, A.H., Wilcox, B.A., & Hirth, J.P., (1970). Dislocation substructure induced by creep in molybdenum single crystals. *Acta Metallurgica*, V.18, pp.381-397, ISSN 0956-7151.

Hay, D.R., Scogerboe, R.K., & Scala, E. (1968). Electron beam zone purification and analyses of tungsten. *Journal of Less-Common Metals*, Vol.15, No.2, pp.121-127, ISSN 0022-5088.

Hurle, D.T. (1977). Control of diameter in Czochralski and related crystal growth techniques. *Journal of Crystal Growth*, Vol.42, pp.473-482, ISSN 0022-0248.

Katoh, M., Iida, S., Sugita, Y., & Okamoto, K. (1991). X-ray characterization of tungsten single crystals grown by secondary recrystallization method. *Journal of Crystal Growth*, Vol. 112, pp.368-372, ISSN 0022-0248.

Kittel, C. (1996). *Introduction to Solid State Physics* (7th edition), Wiley, New York, USA.

Kobayashi, N. (1977). Power required to form a floating zone and the zone shape. *Journal of Crystal Growth*, Vol.43, pp.417-424, ISSN 0022-0248.

Kobayshi, N., & Wilcox, W.C. (1982). Computational studies of convection in a cylindrical floating zone. *Journal of Crystal Growth*, Vol.59, pp.616-624, ISSN 0022-0248.

Langer, J.S. (1980) Instabilities and pattern formation in crystal growth. *Revue of Modern Physics*, Vol.52, No.1, pp.1-28, ISSN 0034-6861.

Liu, J., & Zee, H.R. (1996). Growth of molybdenum-based alloy single crystals using electron beam zone melting. *Journal of Crystal Growth*, Vol.163, pp.259-265, ISSN 0022-0248.

Markin, S.N., Ermolov, S.N., Sasaki, M., van Welzenis, R., Stinov, E.D., Glebovsky, V.G., & Brongersma, H.H. (2006). Scattering of low-energy ions from the surface of a W(211) single crystal. *Physics of Metals & Metallography*, Vol.102, No.3, pp.274-278, ISSN 0031-918X.

Markin, S.N., Ermolov, S.N., Sasaki, M., van Welzenis, R.G., Glebovsky, V.G., & Brongersma, H.H. (2010). On a peculiarity of low-energy ion scattering from well-ordered *bcc* W(211) surface. *Nuclear Instruments and Methods in Physics Research B.*, Vol.B268, pp.2433-2436, ISSN 0168-9002.

Moest, B., Glebovsky, V.G., Brongersma, H.H., Bergmans, R.H., Denier van der Gon, A.W., & Semenov, V.N. (1998). Study of Pd single crystals grown by crucibleless zone melting. *Journal of Crystal Growth*, Vol.192, pp.410-416, ISSN 0022-0248.

Mullins, W.W., & Sekerka R.F. (1964). Stability of a planar interface during solidification of a dilute binary alloy. *Journal of Applied Physics*, Vol.35, No.2, pp.444-451, ISSN 0021-8979.

Mundy, J.N., Rotman, S.J., Lam, N.Q., Hoff, H.A., & Nowicki, L.J. (1978). Self-diffusion in tungsten. *Physical Revue*, V.B12, No.12, pp.6566-6575, ISSN 1098-0121.

Murphy, Y.A. (1987). Numerical simulation of flow, heat and mass transfer in a floating zone at a high rotational Reynolds numbers. *Journal of Crystal Growth*, Vol.83, No.1, pp.23-34, ISSN 0022-0248.

Nes, E., & Most, W. (1966). Dislocation densities in slow cooled aluminum single crystals. *Philosophical Magazine*, Vol.13, No.124, pp.855-859, ISSN 1478-6435.

Otani, S., Tanaka, T., & Ishizawa, Y. (1984). Temperature distribution in crystal rods with high melting points prepared by a floating zone technique. *Journal of Crystal Growth*, Vol.66, No.2, pp.419-425, ISSN 0022-0248.

Pfann, W.G. (1966). *Zone Melting* (2nd edition), Wiley, New-York, USA.

Reid, C.N. (1966). Dislocation widths in anisotropic *bcc*-crystals. *Acta Metallurgica*, V.14, pp.13-16, ISSN 0956-7151.

Riedle, J., Gumbsh, P., Fishmeister, Glebovsky, V.G., & Semenov, V.N. (1996). Cleavage fracture and the brittle-to-ductile transition of tungsten single crystals. *Fracture - Instability Dynamics, Scaling, and Ductile/Brittle Behavior*, R.L. Blumberg, J.J. Mecholsky, A.E. Carlsson & E.R. Fuller, eds., Vol.409, pp.23-28, MRS.

Riedle, J., Gumbsh, P., Fishmeister, Glebovsky, V.G., & Semenov, V.N. (1994). Fracture studies of tungsten single crystals. *Materials Letters*, Vol.20, (August 1994), pp.311-317, ISSN 0167-577X.

Shah, J.S., & Wills, H.H. (1975). Zone melting and applied techniques. In: *Crystal Growth*, B.R. Pamplin, ed., Pergamon Press, London, New York, p.194.

Shah, J.S. (1980). Zone refining and its applications (2nd edition). In: *Crystal Growth*, B.R. Pamplin, ed., pp.301-355, Pergamon Press, Oxford.

Surek, T., & Chalmers, B. (1975). The direction of the surface of crystal in contact with its melt. *Journal of Crystal Growth*, Vol.29, No.1, pp.1-11, ISSN 0022-0248.

Shipilevsky, B.M., & Glebovsky, V.G. (1989). Competition of bulk and surface processes in the kinetics of hydrogen and nitrogen evolution from metals into vacuum. *Surface Science*, Vol.216, pp.509-527, ISSN 0039-6028.

Wilcox, W.R., & Fuller, L.D. (1965). Turbulent free convection in Czochralski crystal growth. *Journal of Applied Physics*, Vol.36, pp.2201-2205, ISSN 0021-8979.

Yadava, R.D.S., Sinha, S., Sharma, B.B., & Warrier, R. (1985). Grain growth mechanism during recrystallization of a tellurium rich $Hg_{1-x}Cd_xTe$ cast. *Journal of Crystal Growth*, Vol.73, No.2, pp.343-349, ISSN 0022-0248.

4

Application of Orientation Mapping in TEM and SEM for Study of Microstructural Evolution During Annealing – Example: Aluminum Alloy with Bimodal Particle Distribution

K. Sztwiertnia, M. Bieda and A. Kornewa
Polish Academy of Sciences, Institute of Metallurgy and Materials Science, Krakow,
Poland

1. Introduction

There are still considerable gaps in the understanding of the recrystallization processes of metallic materials, which reduce the possibility of controlling their course and introducing technological modifications aimed at obtaining desirable properties. The lack of a complete explanation can be attributed to the high complexity of the phenomenon, which consists of a superposition of the local nucleation and grain growth processes. These processes depend strongly on the characteristics of the matrix, which is typically complex and heterogeneously deformed. The phenomenology of the process and its energetic causes are known because they were examined long ago, e.g., (Humphreys & Hatherly, 2002). On the other hand, the relevant physical mechanisms that control the nucleation and growth of new grains are not entirely clear. This uncertainty exists, among other reasons, because the origin of the crystallographic orientations of the nuclei is usually not known.

1.1 Crystallographic orientation and orientation characteristics of materials

The crystallographic orientation is a feature of a material that is defined at any point of the sample at which the ordering of the crystal lattice is not disturbed (or not significantly disturbed). It can be generally said that almost all of the basic quantities that characterize a polycrystalline material and its properties have a direct or complex relationship to the orientation (g_i), which is a function of the coordinates x_i, y_i, z_i of a point in the sample. The $g(x,y)$ function described in a plane of the sample defines the orientation topography (commonly called the orientation map). According to the definition of the term, the orientation at any point (x_i, y_i) in the sample is given by a rotation that brings the local sample reference system with its origin at the point (x_i, y_i) into coincidence with the crystal reference system. The orientation is described unambiguously by three parameters, which can be expressed in different ways. Usually, for the convenience of calculation, the Euler angles φ_1, Φ, φ_2 are applied. In some cases, the parameters of the rotation axis (θ, ψ) and the rotation angle ω are used because this scheme is easy to visualize. If the orientation is described by a greater number of parameters, then the parameters depend on each other.

Such is the situation in the case of crystallographic indices {hkl}<uvw>, commonly used in practice. Extensive analysis of orientation problems can be found in (Morawiec, 2004). If an orientation map is obtained, the grains or subgrains may be reconstructed by identifying areas whose pixels have orientations within a specified range. The knowledge of the orientation topography enables the identification of grain and subgrain boundaries, as well as other microstructural inhomogeneities, by selection of misorientations between neighboring measuring points. This approach enables stereological analysis with regard to the crystallographic orientation. Based on the orientation topography, the orientation characteristics of the microstructure can be determined (Pospiech et al., 1993). The set of orientation characteristics comprises the "principal distributions", which are texture functions determined by the whole set of measurements and "partial distributions", in which only part of measurements are needed. The most important of the principal distributions is the well-known orientation distribution function (ODF), which describes the crystallographic texture of a material. The ODF is defined by the density of the global orientation distribution of the grains (taking into account their volume fraction). Another "principal distribution" is the orientation difference distribution function (ODDF). The ODDF contains all possible misorientations between the measured orientations. To investigate local textures (or microtextures) in selected areas of inhomogeneities or in the environment of preferred orientations, partial distribution functions are applied. The most important of these functions is the misorientation distribution function (MODF), which describes the distribution of misorientation between the nearest neighbor grains. The other partial orientation distributions are statistical quantities related to orientation and misorientation, which may be related to the properties of a material, its anisotropy or specific stages through which material passes (Pospiech et al., 1993).

Many of the essential properties of polycrystalline materials and their anisotropy depend directly or indirectly on the topographical arrangement of orientations. Using the orientation topography, the material properties that depend on the character and distribution of the grain boundaries can be described, for example, segregation or corrosion. The knowledge of the orientation topography is of basic importance for the understanding of many processes that occur in the material, such as deformation, recrystallization, phase transformation or diffusion.

It is, therefore, not surprising that the characterization of the microstructure based on the sets of measured orientations has advanced as a well-established technique, known as Orientation Microscopy (OM). The main concept behind this technique is the automatic collection and indexing of many electron diffraction patterns that are correlated with sample coordinates. The development of new generations of computer-controlled electron microscopes has improved their spatial resolution and increased the rate at which large sets of Electron Back Scattered Diffraction (EBSD) patterns can be collected and processed in a Scanning Electron Microscope (SEM), e.g., (Dingley, 1984; Wright & Adams, 1992; Adams & Dingley, 1994; Schwartz et al., 2009). Systems created orientation topographies using EBSD in SEM are now very common, and commercially available versions of this technology are essentially fully automated, e.g., (HKL, 2007; TSL, 2007).

1.2 Orientation imaging microscopy for recrystallization study

Obviously, the analysis and modeling of the recrystallization of a deformed metallic material requires description of the microstructure evolution during annealing that is as

Application of Orientation Mapping in TEM and SEM for Study of Microstructural Evolution During
Annealing – Example: Aluminum Alloy with Bimodal Particle Distribution

73

complete as possible. Such a description may be based on the orientation topographies obtained by OM techniques in systematic measurements of a sample that undergoes a specific deformation and annealing process, e.g., (Zaefferer et al., 2001; Sztwiertnia, 2008).

To study the recrystallization (particularly its early stages), a high spatial resolution in the orientation measurement is required. Unfortunately, the spatial resolution that can be achieved by EBSD/SEM measurements is relatively low. It falls approximately one order of magnitude behind the spatial resolution in conventional SEM imaging, and still further behind when compared to the spatial resolution of a transmission electron microscope (TEM). The measurement is limited in this way because the inherent resolution of EBSD is governed not by the diameter of the beam spot at the point of impact on the surface, but primarily by the excitation volume. This quantity is the fraction of the interaction volume of the primary electrons within the sample from which the pattern-forming electrons are back diffracted and leave the crystal, without further scattering. This volume is strongly dependent on the type of electron gun and the material being investigated. Tilting of the sample during the EBSD measurement further degrades the spatial resolution and produces resolution along the beam direction on the sample surface that is approximately three times worse than the resolution along the direction perpendicular to the beam, e.g., (Schwartz et al., 2009). For these reasons, the best achievable spatial resolution in EBSD/SEM special cases is in the order of approximately 30 nm; however, the practical limit is approximately 100 nm. This limitation restricts the utility of EBSD/SEM for the investigation of very fine-grained and deformed microstructures, as in the case of the early stages of recrystallization. To obtain better spatial and angular resolution, similar systems have been developed for TEM, e.g., (Haessner et al., 1983; Dingley, 2006; Morawiec, 1999; Morawiec et al., 2002; Rauch & Dupuy, 2005). Despite some restrictions, such as the currently unsolved problems of image analysis, difficulties in measurement automation and sample preparation, the OIM technique applied to TEM offers spatial resolution better than 10 nm and can be used for quantitative analysis of structures at the nanoscale. Such a system, built at the Institute of Metallurgy and Materials Science (Morawiec et al., 2002; Sztwiertnia et al., 2006; Bieda-Niemiec, 2007), was used for a study of the recrystallization of 6013 aluminum alloy (Sztwiertnia et al., 2007; Bieda et al., 2010).

The 6013 aluminum alloy was chosen as a prototype material that represents a group of commercial alloys with a bimodal second phase particle distribution. The second-phase particles are used to control the strengthening, grain size and texture of the alloy. Such alloys can be interesting for examination of the role of the second phase particles in the recrystallization process, e.g., (Humphreys & Hatherly, 2002; Sztwiertnia et al., 2005; Ardakani & Humphreys, 1994).

To elucidate the mechanisms of the alloy microstructure transformation during annealing, *in situ* TEM experiments and combined calorimetric–microscopic investigations of bulk samples were carried out. The *in situ* experiments were necessary to provide information about the temporal relationships between changes that occur in the metal at the beginning of recrystallization. Dynamic studies using SEM or TEM should be capable of providing the required information. Because of its already mentioned limitations, the EBSD/SEM measurements of localized strain in the deformed polycrystal, which are particularly interesting as nucleation sites, give rather poor information about the orientation. For the *in situ* studies of such regions, measurements using convergent beam electron diffraction

(CBED) or microdiffraction in a TEM are more suitable, although the proximity of the free surface in the thin foils can be a complicating factor during annealing. The first *in-situ* TEM observations, obtained by Bailey in 1960 and Hu in 1963, indicated differences in the recrystallization processes that can occur during the annealing of bulk samples and thin foils. As a consequence, many researchers have been skeptical about the results of such experiments up to now. However, the results obtained later by other authors e.g., (Roberts & Lehtinen, 1972; Hutchinson & Ray, 1973; Sztwiertnia & Haessner, 1994) allow the determination of the experimental conditions, which ensures that the changes directly observed in an annealed foil are at least similar to those occurring in a bulk sample. In general, recrystallization is easiest in orthogonal sections from a rolled sheet in which the grain boundaries, extending from top to bottom of the foil, present the most favorable distribution of driving potential for migration. Grooving grains are not strongly inhibited in the thin foil regions and frequently extend nearly to the edge of the foil (Hutchinson & Ray, 1973). Nevertheless, because of the thermal grooving, the recrystallization front always stops in foil thinner than a certain critical value, which approximately depends on the fineness of its microstructure. One can increase the usable foil thickness by increasing the accelerating voltage. The impact of the sample thickness on grain boundary movement explained in greater detail by Roberts & Lethinen, 1972.

In highly deformed 6013 aluminum alloy, the critical thickness for foils cut from planes perpendicular to the sheet is so low that it allows *in situ* observation of nucleation events (and to some extend the growth of the nuclei) to be carried out in a conventional TEM operated at 200 kV (Sztwiertnia et al., 2005; Sztwiertnia et al., 2007; Bieda et al., 2010). To examine the significance of the *in situ* experiments, the thin foils annealed in the TEM were compared to thin foils prepared from bulk samples heated in a calorimeter to obtain a specified recrystallization stage. The comparison shows that the processes occurring in both types of foils were at least qualitatively the same.

2. Example: Recrystallization of aluminum alloy with bimodal particle distribution

2.1 Material and investigation methodology

The changes of the microstructure during annealing were examined in the case of the polycrystalline aluminum alloy 6013 (Table 1), which was previously identified as the prototype of materials with a bimodal precipitate distribution.

Mg	Si	Cu	Mn	Fe	others	Al
1.15	1.0	1.1	0.3	0.5	0.15	remainder

Table 1. 6013 aluminum alloy chemical composition (% wt.).

Samples for testing were supersaturated, then aged and reversibly cold-rolled up to the 75 and 90 % of the 10 mm value. The deformed samples were examined by means of non-isothermal annealing in a differential calorimeter (DC). It was found that the spectrum of released stored energy contained several peaks. Next, a new series of samples were heated in the calorimeter to the selected temperatures, rapidly cooled, and then analyzed in the TEM. The tests in the TEM were complemented by SEM examinations. Because of the poor quality of the orientation

Application of Orientation Mapping in TEM and SEM for Study of Microstructural Evolution During
Annealing – Example: Aluminum Alloy with Bimodal Particle Distribution

75

topographies measured by the standard EBSD/FEG/SEM techniques, these measurements were used only to detect the presence of newly recrystallized grains (that is, only those diffraction patterns were taken into account that had a high image quality and showed local areas with low dislocation densities). The high degree of deformation also resulted in low quality TEM diffraction patterns. However, it was still possible to measure enough single orientations in the TEM to construct orientation topographies for all of the microstructural elements of the cold-rolled material. The investigations also included a comparison of the deformed alloy microstructure with that of the pure metal and the investigation of the recrystallization process dynamic. The latter investigation consisted of the *in situ* measurements in the TEM. This measurement was necessary to obtain information about the time sequence of changes occurring in the material at the beginning of the recrystallization. In the investigated material, the sequence of events that occurs in the deformation zones is of particular interest because such areas undergo intense nucleation.

2.2 Deformation state

First, the deformation microstructure of the alloy was compared with that of the pure metal. The microstructure of commercially available pure aluminum (3N Al), reversibly cold rolled to 90%, has been chosen for comparison. Both microstructures are built of elongated in the rolling direction (RD) and lie nearly parallel to the sheet plane grains and subgrains, as shown in Fig. 1 b and 2 a. Some significant differences became evident when the measured orientation topographies were compared. In the alloy matrix deformed to 75 %, the distances between the high angle grain boundaries (HAGB) in the normal direction (ND) to the sheet plane were typically smaller than 1 µm, as shown in Fig. 2 c. In the pure metal, the thickness of similarly oriented layers often exceeded 10 µm, even for the deformation of 90 %, as shown in Fig. 1 a, c. These layers were composed of parallel bands or clusters of subgrains that were strongly elongated in the RD. In some of them, relatively small but accumulating disorientation angles[1] occurred (Fig. 1c). Because of this accumulation, large disorientation angles (up to ~ 20 °) between the first and the last subgrain in the band occurred frequently. Orientation changes of this type are characteristic for transition bands (Dillamore et al., 1972). The alternation of orientation patterns was also recognized. The high frequency of the low angle grain boundaries (LAGB) inside the band indicates a well-developed subgrain structure. With no further analysis of the pure metal deformation microstructure, we can only conclude that it consists of thick deformation and transition bands, and the density of HAGBs in the ND is low.

The crystallographic orientation characteristics of the alloy were quite different from those of the pure metal. The matrix consisted of well-developed HAGBs. The distances between them along ND were much smaller than those in the pure metal and never exceed a few hundreds of nanometers. LAGBs in the elongated thin matrix grains were less ordered and created a less-expanded subgrain structure. In the laminar microstructure of the alloy, small (<<1 µm) and large (1÷3 µm) precipitate particles of the second phase were scattered. Around the large particles, zones of localized strain were identified. These deformation zones consisted of

---

[1]A given misorientation can be described by a rotation axis and an angle of rotation. For a material with crystal symmetry, there is more than one angle of rotation. The angle with the absolute value smallest of all possible angles of rotation is called the disorientation angle.

ultrafine (50 – 200 nm) grains and more or less bent microbands of the matrix (Fig. 3 b, 4 b). The orientation image of the alloy clearly suggests that microstructural evolution occurs by grain subdivision at a very small scale compared to the original grain size, which was approximately 100 µm in this case. This result is presented in Fig. 2 c and 3 d, showing large orientation variations over a region as small as a few micrometers. Such rapid changes in orientation may demonstrate that volumes characterized by a combination of slip systems can be very small.

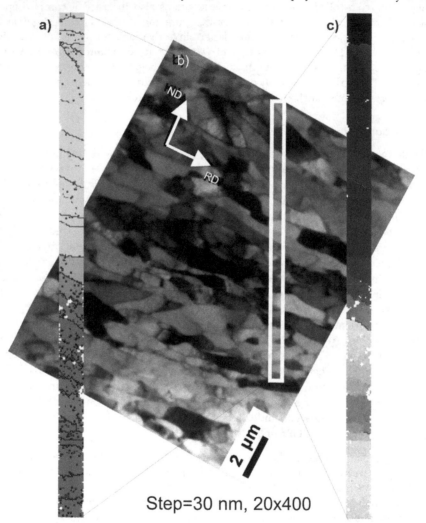

Step=30 nm, 20x400

Fig. 1. As-deformed microstucture of 90% cold-rolled aluminum, the TEM bright field image (b) and orientation topographies (a and c). On the map (c) the color change indicates a deviation from the initial orientation (dark blue) to the disoriented orientation (light blue); black lines (a) - low angle grain boundaries (>1 º), red lines (c) - the boundaries with a disorientation angle > 5 °; the points where the diffraction pattern has not been solved are shown in white.

Application of Orientation Mapping in TEM and SEM for Study of Microstructural Evolution During
Annealing – Example: Aluminum Alloy with Bimodal Particle Distribution

77

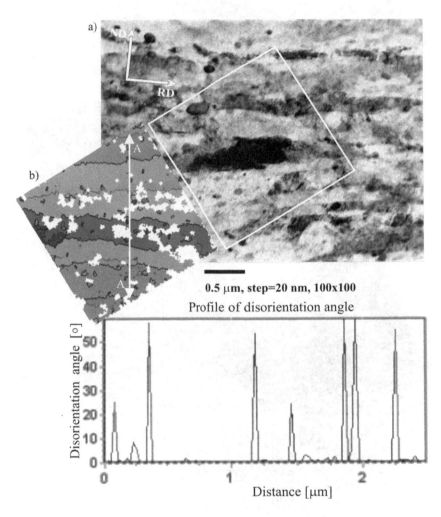

Fig. 2. a) As-deformed microstucture of 75% cold-rolled 6013 alloy, longitudinal section,
TEM. b) Orientation topography of the matrix area, black lines show high angle grain
boundaries; white regions are not indexed. c) Disorientation angle profile along A-A line (b),
(Sztwiertnia et al., 2007).

The global crystallographic textures of both materials were similar and corresponded to the
well-known rolling texture of FCC metals with high stacking fault energies (such as pure
aluminum). Such a texture is characterized by the concentration of components along two
orientation fibers. The main one, called the α fiber, runs diagonally through the orientation
space containing the preferred texture components S {123}<634> and Cu {112}<111>. The
other one, called the β fiber, includes Goss {011}<100> and Bs {011}<211> components
(Sztwiertnia et al., 2005).

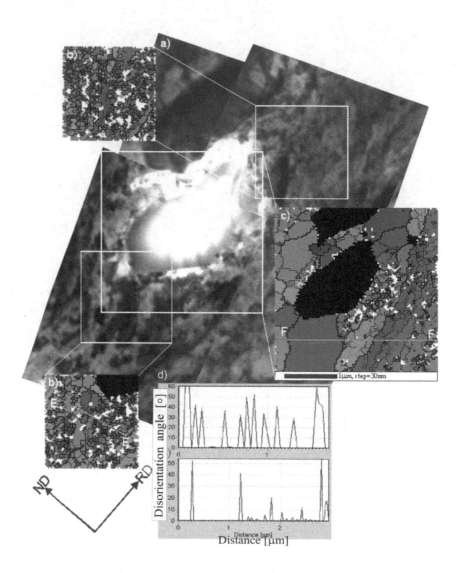

Fig. 3. a) As-deformed microstucture of 75% cold-rolled 6013 alloy, the deformation zone around the large particle, longitudinal section, TEM. Orientation topographies in areas of the deformation zone before (b) and after (c) heating *in situ* in TEM; particles of the second phase are shown in black, white regions are not indexed; thick lines indicate high angle grain boundaries, thin lines indicate low angles grain boundaries. d) Examples of disorientation angle profiles along E-E and F-F (Sztwiertnia et al., 2007).

Application of Orientation Mapping in TEM and SEM for Study of Microstructural Evolution During
Annealing – Example: Aluminum Alloy with Bimodal Particle Distribution

79

2.3 Microstructure changes during annealing

The cold-rolled alloy was tested using the non-isothermal annealing method in a differential calorimeter. On the basis of these tests, as well as the microscopic analysis of the microstructures of the appropriately annealed samples, it was possible to state that the two separate peaks of the stored energy release correspond to the two stages of the recrystallization process (Fig. 6 a).

The *in situ* tests in the TEM allowed the determination of the sequence of events occurring at the beginning of recrystallization in the deformation zones around the large particles and in the matrix beyond those areas. First, the deformation microstructure was carefully examined. In the deformation zones around the large second phase particles, small grains and distorted fragments of microbands were identified. The small grains were approximately 50-200 nm in size. The strong orientation changes, greater than 15°, either identify HAGBs lying at distances lower than 200 nm or they are an effect of strong grain bending, as shown in Fig. 3 b, 4 b. To precisely distinguish between these two phenomena, small-step orientation maps were generated. The broad distribution of orientations in the deformation zones tended to group in the range of the deformation components after rotation around the transverse (TD) or the normal direction (ND) to the sheet plane (Sztwiertnia et al., 2005). This type of rotations suggests that at least a part of the strong orientation changes may be a result of the accumulation of small disorientations along a bent grain. In the matrix outside of the DZs, the HAGBs lie roughly parallel to the sheet plane at distances of 0.5 - 1 µm along the ND (Fig. 2 b).

After the sample was annealed in the microscope, the same areas as in the deformation state were investigated. Figures 3 c and 4 c show examples of DZ orientation maps after *in situ* annealing. Nuclei and new grains appeared in the vicinity of the deformation zone. The orientations of crystallites in the deformed state commonly lay in the area of a particular new grain or nucleus (Fig. 4 d, e). For each orientation of a new grain, at least one similarly oriented fragment of the deformed matrix was found. Some of the nuclei were growing within the zone defined by the migration of HAGBs, a result that was confirmed by a significant reduction of their density in those zones after annealing to the temperature of the first peak (Fig. 3 c, d and 4 b, c). The shape of some new grains suggests that they could have been formed as a result of the local recovery of strongly bent fragments of matrix microbands, as shown in Fig 4 c.

Partial misorientation distribution functions (PMDF) were calculated between the grains in the deformed state and the new grains that appeared at the same location (Fig. 5). PMDFs for both the 75 and 90 % deformed materials show a random distribution of misorientations. This distribution suggests that there was no special orientation relationship describing favored growth.

In the deformation zones at very early stages of the recrystallization, broad-spectrum HAGBs appear to have been active and mobile. At that time, no migration of HAGBs in the matrix areas outside the zones was observed.

To examine the significance of the *in situ* experiments, the microstructures of TEM-annealed thin foils were compared with the microstructures of thin foils prepared from bulk samples heated in a calorimeter to obtain a specified recrystallization stage. The comparison shows that the processes occurring in the thin foils and in the bulk samples were, at least qualitatively, the same. The temperature range of the first recrystallization peak was found to produce nucleation, which is accompanied by some limited enlargement of new grains in the sheet plane.

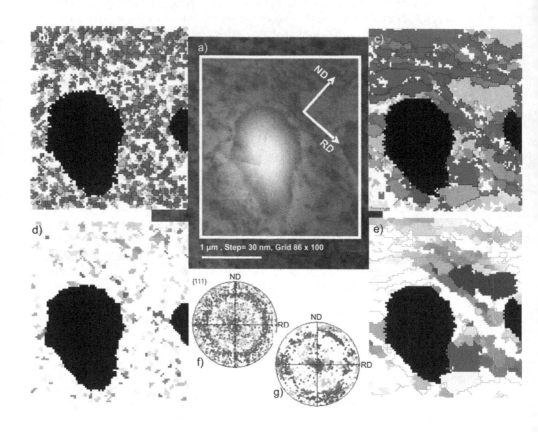

Fig. 4. a) Microstructure of 6013 aluminum alloy cold rolled to 90%, the deformation zone surrounding a large particle (a), longitudinal section, TEM. Orientation topographies and pole figures in areas of the deformation zone before (b, f) and after (c, g) heating *in situ* in TEM; particles of the second phase are shown in black, white regions are not indexed; thick lines indicate high angle grain boundaries, thin lines indicate low angle grain boundaries. Areas of similar orientations (blue, red, yellow) before and after annealing (d, e) (Bieda et al., 2010).

Application of Orientation Mapping in TEM and SEM for Study of Microstructural Evolution During
Annealing – Example: Aluminum Alloy with Bimodal Particle Distribution

81

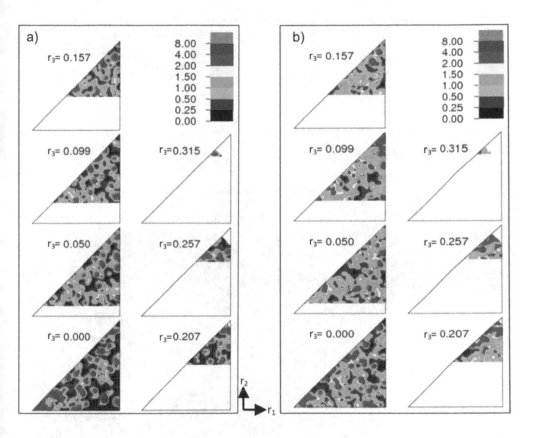

Fig. 5. Partial misorientation distribution functions showing the orientation relationships
between the crystallites from the deformation zones (before annealing) and the new grains
appearing in their positions after annealing (a) 75% cold-rolled 6013 aluminum alloy, (b)
90% cold-rolled 6013 aluminum alloy; Rodrigues' representation r_1, r_2, r_3, cross-section
r_3=const., asymmetric domain (O, O) (Bieda et al., 2010).

By way of example, Fig. 6 c shows the orientation topography in the 75 % cold-rolled sample that was heated to 330 °C in a calorimeter. The appearance of new grains with sizes up to a few micrometers can be observed around the large particle. These new grains, similar to those in the *in situ* experiment, were formed as a result of nucleation and the consequent growth of nuclei inside the zone. At the examined temperature, the arrangement of LAGBs can also be observed in the elongated grains of the matrix outside the zone. The annihilation of these boundaries takes place at higher temperatures, near the end of the first peak.

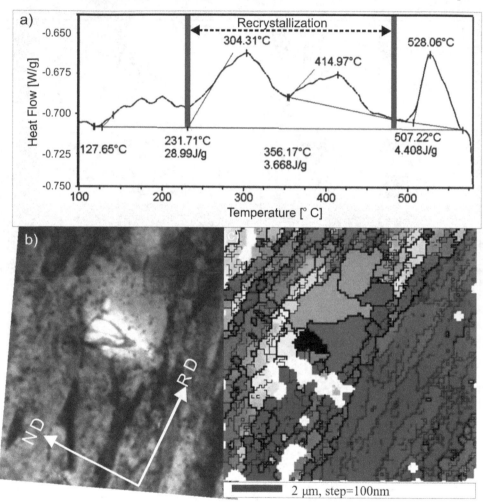

Fig. 6. a) Power differences, representing the release of stored energy from 75 % cold-rolled 6013 aluminum alloy as a function of annealing temperature. b, c) As-deformed microstructure and orientation topography of 6013 aluminum alloy 75 % cold-rolled and subsequently heated in the calorimeter to 330 °C: particles of the second phase are shown in black, white regions are not indexed; thick lines indicate high angle grain boundaries, thin lines indicate low angle grain boundaries; TEM, (Sztwiertnia et al., 2007).

Application of Orientation Mapping in TEM and SEM for Study of Microstructural Evolution During
Annealing – Example: Aluminum Alloy with Bimodal Particle Distribution

83

The phenomenon of the annihilation of LAGBs and the growth of new grains in the sheet plane (mainly, parallel to the RD) was also observed in the SEM results (Fig. 7). The SEM measurements of the orientation topographies were made on samples heated to appropriate temperatures from the range of the first peak (Fig. 5 a). These observations fully confirm the TEM results and show that after the recrystallization of the deformation zones, local recovery processes (coalescence of subgrains) take place in the matrix. These local recovery processes lead to the annihilation of the LAGBs between chains of subgrains lying parallel to the sheet plane and, consequently, to the production of long grains with a low density of lattice defects. The growth of the elongated grains in the ND occurs rarely or not at all within the temperature range of the first peak. The migration of HAGBs in the matrix becomes the main process within the temperature range of the second peak. Heating of the sample to the temperature of the end of this peak leads to the complete discontinuous recrystallization of the material (Fig. 7 c). The recrystallized microstructure is dominated by elongated grains (up to 30 μm in length along the RD). It also includes groups of smaller grains, which are often almost equiaxial. After the discontinuous recrystallization, the grains are a few times thicker than they are in the state observed at the end of the first stage.

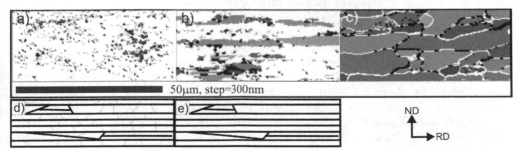

Fig. 7. Orientation topographies of recrystallized grains in 6013 aluminum alloy 75% cold rolled and subsequently heated in the calorimeter to: (a) 330 °C, (b) 350 °C and (c) 480 °C, regions of unsolved diffraction (approximately corresponding to the deformed areas) are shown in white, EBSD/SEM/FEG. d) and e) Schematic representation of the deformed microstructure before and after annealing to the temperatures from the end of the first recrystallization peak, Fig. 2; thick lines indicate high angle grain boundaries, thin lines indicate low angle grain boundaries.

3. Summary

Orientation topography provides basic local and global information about microstructures by allowing the identification and description of occurring regularities. The OM technique in TEM is a useful tool for the quantitative and qualitative characterization of fine crystalline and deformed microstructures in polycrystalline materials. It is possible to obtain information about grain distribution, misorientation between grains, material phases, the local orientation distribution function, and the misorientation distribution function. Replacement of the SEM measurements by TEM measurements improves the spatial resolution to a few nanometers. Both SEM and TEM can be used for complementary analysis of crystalline materials at the "micro" and "nano" scale, respectively. Together with *in situ* studies, orientation mapping in TEM can provide additional information about the behavior of a material during annealing, particularly in zones of greater deformation. The example of orientation characteristics presented in this chapter illustrates only some aspects of the applicability of these techniques. Orientation mapping in a transmission electron microscope was successfully applied to the study of microstructural changes during the initial stage of recrystallization in an aluminum alloy with a bimodal second-phase particle distribution. The images of the microstructure in the representative areas of a sample of deformed aluminum alloy 6013, described by measurements of orientation topography, shows greatly advanced grain fragmentation.

In situ investigations in TEM, calorimetric measurements, and orientation mapping in TEM and SEM demonstrate that the recrystallization of the tested material can be considered to occur as a number of partly overlapping processes that proceed in two stages. These stages correspond to the two separate stored energy release peaks. In the initial stage, the deformation zones around large second phase particles act as sites for particle-stimulated nucleation. This nucleation is accompanied by the growth of nuclei. However, the migration of high angle grain boundaries only occurs in deformation zones. At the same range of temperatures, some enlargement of new grains in the matrix (outside of the deformation zones) was also observed. The formation of grains elongated primarily in the direction parallel to the rolling direction may be correlated to the processes of local recovery, which is triggered in the deformation zones. Grain elongation then continues to develop along the bands of the deformed matrix in the directions of low orientation gradients. The elongated grains appear because of the annihilation of low angle grain boundaries between chains of subgrains lying in layers parallel to the sheet plane. As a consequence, new grains often have a plate-like character, with their shorter axis parallel to the sheet plane normal direction. Their lengths along the rolling direction may exceed 50 µm, while their thickness corresponds approximately to the distance between high angle grain boundaries in the normal direction outside the deformation zones; this dimension was not observed to exceed a few micrometers. In the second stage, high angle grain boundaries were observed to migrate in the direction of the high orientation gradient. This migration, mostly in the normal direction, was limited to "free areas" of the deformed matrix between bands of new grain formation in the initial stage of recrystallization.

4. References

Adams B.L., Dingley D.J (1994), *Orientation Imaging Microscopy: New Possibilities for Microstructural Investigations using Automated BKD Analysis*, Mater. Sci. Forum, 157-62 31.

Ardakani M.G., Humphreys F.J. (1994), *The annealing behavior of deformed particle-containing aluminum single crystals.* Acta Metal. Mater., 42, 763.

Bailey J.E. (1960), *Electron microscope observations on the annealing process occurring in cold-worked silver*, Phil. Mag., 5, 833.

Bieda M.,.Sztwiertnia K, Korneva A., Czeppe T., Orlicki R. (2010), *Orientation mapping study on the inhomogeneous microstructure evolution during annealing of 6013 aluminum alloy*, Solid State Phenom, 16, 13.

Bieda-Niemiec M. (2007), *Opracowanie systemu do automatycznego pomiaru map orientacji w transmisyjnym mikroskopie elektronowym do analizy mikrostruktury drobnoziarnistych materiałów metalicznych*, Thesis, Kraków IMIM PAN, in polish.

Dillamore I.L., Morris P.L., Smith C.J.F., Hutchinson W.B. (1972), *Transition Bands and Recrystallization in Metals,* Proc. Roy. Soc., 329A, 405.

Dingley D.J. (1984), *On-line determination of crystal orientation and texture determination in SEM,* Proc. Roy. Microsc. Soc., 19, 74.

Dingley D.J. (2006), *Orientation Imaging Microscopy for the Transmission Electron Microscope,* Mikrochim. Acta, 155, 19.

Haessner, F., Pospiech, J. and Sztwiertnia, K. (1983), *Spatial arrangement of orientations in rolled copper,* Mat. Sci. Eng.1, 1.

HKL (2007) http://www.oxford-instruments.com

Hu H. (1963), *Electron Microscopy and Strength of Crystals,* Interscience, London, 564.

Humphreys F.J. and Hatherly M. (2002), *Recrystallization and Related Annealing Phenomena,* Pergamon Press, Oxford.

Hutchinson W.B., Ray R.K. (1973), *On the feasibility of in situ observations of recrystallization in the high voltage microscope,* Phil. Mag., 28, 953.

Morawiec A. (1999), *Automatic orientation determination from Kikuchi patterns,* J. Appl. Cryst. 32, 788.

Morawiec A. (2004), *Orientations and Rotations. Computations in Crystallographic Textures.* Berlin, Heidelberg, New York: Springer-Verlag.

Morawiec A., Fundenberger J.J., Bouzy E., Lecomte J.S. (2002), *EP-a program for determination of crystallite orientations from TEM Kikuchi and CBED diffraction patterns"* J. Appl. Cryst., 35, 287.

Pospiech J., Lücke K. and Sztwiertnia K. (1993), *Orientation Distribution and Orientation Correlation Functions for Description of Microstructures,* Acta Metall. Mater. , 41, 305.

Rauch E.F., Dupuy L. (2005), *Rapid spot diffraction patterns identification through template matching,* Arch. Metall. Mater., 50, 87.

Roberts W., Lehtinen B. (1972), *On the feasibility of in situ observations of recrystallization in the high voltage electron microscope,* Phil. Mag., 26, 1153.

Schwartz A.J., Kumar M.,. Adams B.L, Field D.P. (2009), *Electron Backscatter Diffraction in Materials Science,* ISBN 978-0-387 88135-2, Springer.

Sztwiertnia K. (2008), *On recrystallization texture formation in polycrystalline fcc alloys with low stacking fault energies,* Int. J. Mater. Res., 99, 178.

Sztwiertnia K., Bieda M., Korneva A., Sawina G. (2007), *Inhomogeneous microstructural evolution during the annealing of 6013 aluminium alloy,* Inżynieria Materiałowa, Vol. 3 XXVIII, 476.

Sztwiertnia K., Bieda M., Sawina G. (2006), *Determination of crystallite orientations using TEM. Examples of measurements,* Arch. Metall. Mater. , 51, 55.

Sztwiertnia K., Haessner F. (1994), *In situ observations of the initial stage of recrystallization of highly rolled phosphous copper,* Mater. Sci. Forum, 157-162, 1069.

Sztwiertnia K., Morgiel J., Bouzy E. (2005), *Deformation Zones and their Behaviour during Annealing in 6013 Aluminim Alloy,* Arch. Metall. Mater. 50, 119.

TSL (2007) http://www.edax.com

Wright S.I., Adams B.L. (1992), *Automatic analysis of electron backscatter diffraction patterns*, Met. Trans. A 23, 759.

Zaefferer S, Baudin T, Penelle R (2001), *A study on the formation mechanisms of the cube recrystallization texture in cold rolled Fe-36% Ni*, Acta Mater. , 49, 1105.

5

Physical Metallurgy and Drawability of Extra Deep Drawing and Interstitial Free Steels

Kumkum Banerjee

Research and Development Department, Tata Steel Ltd., Jamshedpur, India

1. Introduction

The aim of this review is to present the underlying physical metallurgy for the development of aluminium killed (Extra deep drawing--EDD) and interstitial free (IF) steels, their recrystallization texture and its subsequent impact on the formability of these steels. A plethora of literature is available and a number of review articles have appeared previously (**Hutchinson, 1984; Ray et al, 1994**). These contributions dealt broadly with the development of cold rolled and annealed textures till 1994 and since then further advances in research had been made on the subject and the present article is intended to provide the progresses made on the subject till date, while also giving a critical review of the subject as a whole.

The automotive industry aims to reduce the weight of outer-body car panels while maintaining strength, formability and dent resistance. However, conventional high strength sheet steels have insufficient formability to meet the drawing requirements of today's more complex outer-body car panels. In the recent years low and ultra low carbon steels like extra deep drawing aluminum killed, interstitial free, interstitial free high strength and bake hardening steels are known for their formability and are extensively used for the auto bodies.

Texture is an important parameter of steel sheets as it induces plastic anisotropy that can be beneficial to drawability of steels (**Hosford & Backholen, 1966; Lankford et al., 1950; Yoshicla, 1974**). The anisotropy is conveniently measured in terms of r_m-value that is the ratio of true width strain to true thickness strain determined through standard tensile tests. r_m-value varies essentially with respect to rolling direction of the sample. Thus, an average of the r-values is taken as r_m, which is expressed through the expression--$(r_0 + 2r_{45} + r_{90})/4)$ – termed as 'normal anisotropy'--where the subscripts, 0, 45 and 90 refer to the tensile specimens with parallel to, 45º and 90º to the rolling direction of the steel sheet. Isotropic steels have r_m-value around 1 while steels suitable for deep drawing applications should have r_m-value 1.8 (Holie, 2000).

High r_m-values correlate well with good deep drawability (**Lankford et al., 1950**). Good drawability also diminishes the edge splitting tendency during hole- expansion tests (**Klein & Hitchler, 1973**). The favourable texture for good deep drawability is a large fraction of the grains oriented with {111} planes parallel to the plane of a sheet (**Whiteley & Wise, 1962**). To ensure satisfactory drawability in these steels, i. e. to increase the depth of drawing and avoid the crack during deep drawing process and at the same time to make the edge on the top of a drawn cup smooth without the phenomenon of earing, the deep drawing sheet is required to possess high plastic anisotropy, r_m and low normal anisotropy, Δr. In other words, to maximize r_m-value and minimize Δr-value, {111}<112> and {111}<110> components of γ-fiber **(Figure 1)** **(Kestens et al., 1996)** are the ideal crystallographic textures for deep drawing steel, because the correct texture gives the proper orientation of slip system so that the strength in the thickness direction is greater than that in the plane of the sheet. If {100} plane parallels rolling plane, the strength is lowest in the thickness direction of sheet. This, in turn, adversely influences the formability of the sheet. The {111}/{100} intensity ratio is reported to be linearly related to r_m **(Held, 1965)** and can easily be determined using X-ray diffractometer measurements of the (222) and (200) lines.

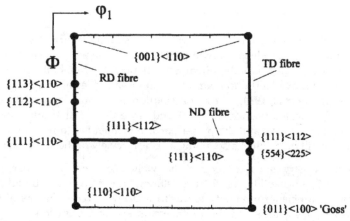

Fig. 1. $\varphi 2=45°$ section of Euler space showing the ideal bcc rolling and recrystallization texture components **(Kestens et. al., 1996)**

2. Recrystallization texture

In several cases, after only small cold deformation, nearly random textures are produced. However, on annealing after very heavy cold reductions, a strong recrystallization texture is usually obtained that may involve the partial retention of the deformation texture but quite often a very different but very strong new texture forms. Thus, the changes in texture that occur during the recrystallization process can be massive while compared to previous texture of the deformed state and in some cases, replaced by an entirely different texture **(Doherty et al., 1997)**.

Two major theories exist for the formation of recrystallization texture-- described as 'oriented nucleation' and 'oriented growth' **(Doherty et al., 1988 & Samajdar, 1994)**. Oriented nucleation is the hypothesis that explains, grains with an orientation that dominates the fully recrystallized texture, nucleate more frequently than do grains of all other orientations. In other words, the oriented nucleation theory assumes orientation selection in the nucleation process based on the orientation dependence of the deformation energy stored in the grains during cold rolling (Tsunoyama, 1998). The high stored energy in (111) grains is considered to be responsible for the (111) orientation nucleation and growth. However, the energy stored in (110) grains is higher than that in (111) grains and another mechanism is required for the preferential development of (111) orientation. To describe the orientation nucleation theory quantitatively, for example for the most discussed case of the formation of 'cube' texture after the recrystallization of heavily rolled FCC metals such as Cu or Al, the fraction of grains, by number, within a selected misorientation, of 10 or 15° from exact cube, α_c, must be normalized by the fraction expected in a random grain structure, α_r (Doherty, 1985). The condition for a strong 'oriented nucleation' effect is that: $\alpha = \alpha_c / \alpha_r$. That is, the frequency of the formation of the new cube grains is much higher than the expected random frequency, so many of the grains will have the special orientation.

While, the oriented growth theory is based on the orientation dependence of the grain boundary mobility (Tsunoyama, 1998). In this theory, orientation relations between recrystallization nuclei and the deformed matrix is responsible for the texture development. However, no significant experimental evidence of oriented growth has been obtained for IF steels until now, even with modern techniques like EBSD (Electron Back-Scattered Diffraction). The oriented growth factor, β, is determined by the relative sizes $\overline{d_c} / \overline{d_r}$ of the cube to the average grains (Doherty, 1985). That is, there is a strong oriented growth effect if: $\beta = \overline{d_c} / \overline{d_r} \gg 1$ (Martin, et al., 1997).In the opinion of Doherty et al (Doherty et al., 1997), the two theories of oriented nucleation and oriented growth should be renamed as: (i) the grain frequency effect; and (ii) the grain size effect, respectively. The reason behind the change was i) the nucleation involves only the growth of a particular subgrain and the terms, oriented nucleation and oriented growth are often taken to indicate specific mechanisms for the frequency or size advantage. Thus, the usage of "frequency" and "size effect" helps avoid such confusion.

The steel recrystallization texture is of major industrial importance. It is found that recrystallization in a cold-worked low carbon steel is mainly controlled by the oriented nucleation theory/grain frequency effect that is governed by the orientational dependence of the stored deformation energy (Hölscher et al, 1991.) The two key recrystallization texture components in steel are {110}<1$\overline{1}$0> and {554}<22$\overline{5}$>. The latter component is just a few degrees away from another recrystallization texture component {111}<112> (Hatherly & Hutchinson, 1979). During recrystallization two major changes take place. The orientation {001}<110> and the orientation spread surrounding the partial α-fiber texture gets eliminated after annealing and some redistribution of intensity in the fiber texture with {111} planes parallel to sheet.

The strength of {111} texture determines the drawability of low carbon and extra low carbon steels. The strength of {111} texture in turn is influenced by chemistry of the steel (Perera et al., 1991; Wilshynsky-Dresler et al., 1995) and the prior technological processing steps, such as hot rolling (Wilshynsky-Dresler et al., 1995 & Perera et al., 1991), cold rolling (Perera et al., 1991) and annealing (Perera et al., 1991, Wilshynsky-Dresler et al., 1995). Many studies have been made on the effect of process conditions and the following principles are obtained for the development of (111) recrystallization texture (Tsunoyama, 1998):

1.   increasing coarseness of precipitates in hot bands;
2.   decreasing grain sizes of hot band;
3.   increasing cold reduction rate;
4.   increasing annealing temperature.

However, these studies are not sufficient to make clear the mechanism of (111) texture development.

Thus, it is important to have a thorough understanding of the underlying physical metallurgy involved so that the desired recrystallization texture is obtained by suitably controlling the prior processing steps to result in formable grade of steels.

3. Processing of aluminium killed EDD steels

In annealed EDD, three types of microstructures are possible depending upon the stage while AlN precipitate forms from Al & N in solid solution (Auburn & Rocquet, 1973):

i.   equiaxed grain structure is obtained while AlN either forms during coiling or after recrystallization during annealing.
ii.   The transition zone recrystallized microstructure forms while recrystallization and AlN precipitation occur simultaneously.
iii. elongated or pancake grain structure is obtained while AlN forms prior to recrystallization during annealing. The AlN precipitates form on the defects of cold rolled structure and thus, recrystallization is retarded. A remarkable enhancement of {111}<uvw> crystallographic orientation occurs as nucleation occurs more rapidly in grains of this type **(Beranger et al., 1996)**

EDD steel sheets are produced by either batch or continuous annealing of cold-rolled steel sheets containing carbon up to about 0.05% and Mn up to about 0.2% **(Sarkar et al., 2004)**. However, the physical processes involved in these processes are different. Thus, batch and continuous annealing processes will be detailed separately in the following sections.

3.1 Hot band texture

The hot band texture of such steels is reported to be nearly random with rotated cube component {001}<110> being approximately 2 times random (2XR) **(Heckler & Granzow, 1970)**.The recrystallization of austenite during hot rolling is reasonably fast and gets completed prior to the transformation to ferrite. Further, in EDD steels no other texture component remains present after hot rolling indicating the fact that the austenite did not have any deformation texture component prior to transformation to ferrite **(Ray et al., 1990)**.

3.2 Cold rolled texture

The cold reduction has an important role in dictating the grain morphology after annealing, texture and mechanical properties. **Figure 2a -2c** depict the effect of cold reduction on grain size, r_m-value (drawability) Δr (planer anisotropy – earing) for an EDD grade steel (C: 0.034%, Mn: 0.21%, Al: 0.06% and N: 0.005%) **(Hebert et al., 1992)**. Thus, an optimized cold reduction must be taken into account while high deep drawability as well as minimum earing is desired.

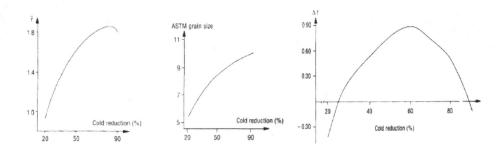

Fig. 2. (a) Variation of $\bar{r}$ (r_m-value) with cold work, (b) variation of grain size with cold work and (c) variation of Δr with cold work for the EDD grain steel (Hebert et al., 1992).

With increasing cold reduction the steel develops both partial α-fiber <110>||RD and γ-fiber {111}||ND. The transformed {001}<110> component also strengthens noticeably. With the increase in cold reduction from 60% to 80%, the strongest texture component shifts from {111}<10> to {112}<110> (**Heckler & Granzow, 1970**)

3.3 Batch annealing

Since the pancake grain structure provides favourable texture for high deep drawability, the processing parameters are set to obtain pancake structure, while transition zone structure is avoided since the latter results coarse grains (ASTM<8) leading to reduced ductility and the risk of orange peel formation during drawing. In order to ascertain that the nitrogen remains in solution, the AlN that forms in the cast material requires to be dissolved during slab reheating (**Auburn & Rocquet, 1973; Meyzaud et al., 1974**). Usually, soaking temperatures of the order of 1200-1250ºC are necessary. In addition, the recombination of Al with N also needs to be prevented during cooling and coiling after hot rolling. To attain this, the finish rolling temperature must be high enough and above Ar3 (**Figure 3**) (**Beranger et al., 1996**) followed by fast cooling in the AlN precipitation range in association with low coiling temperature (<600ºC), **Figure 4** (**Beranger et al., 1996**) to avoid poor ductility and drawability in the annealed steel.

3.4 Continuous annealing

Continuous annealing lines combine several processes including cleaning, annealing, over aging or galvannealing, and sometimes temper rolling, in one continuous operation. In continuous annealing due to high heating rate, recrystallization during annealing occurs at higher temperature than batch annealing and the precipitation of AlN occurs after recrystallization with nitrogen previously in solution. Thus, the nucleation of preferred oriented grains is hindered and due to nitride precipitation, subsequent growth of the recrystallized grains is also restricted. This causes the development of unfavourable texture. Further, the presence of carbides and carbon in solution during recrystallization also assist in the formation of unfavourable texture for drawing.

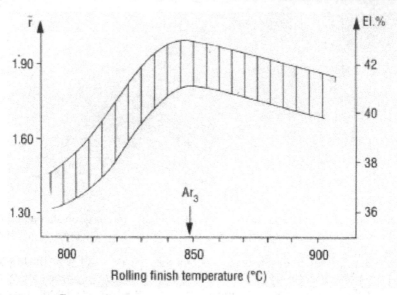

Fig. 3. Variation of $\bar{r}$ (r_m-value) and ductility with finish rolling temperature for a batch annealed EDD steel (**Beranger et al., 1996**).

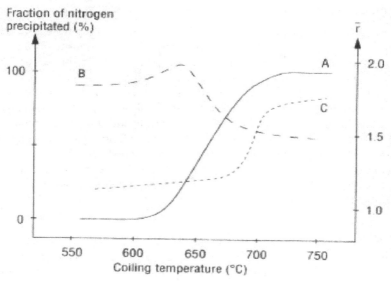

Fig. 4. Variation with coiling temperature of (A) AlN precipitation,(B) the r_m-value in batch annealing and (C) the r_m-value in continuous annealing for a low carbon low manganese EDD steel.(Beranger et al., 1996).

Therefore, in continuous annealing it is endeavoured so that AlN is precipitated prior to annealing, by high temperature coiling after hot rolling. Otherwise, by reheating the as-cast slabs at a temperature too low to take the nitrides back into solution.

However, high temperature coiling has two disadvantages (**Beranger et al., 1996**):

- It causes variation in mechanical properties in the product due to difference in cooling rate between middle and end regions of the coil. A selective coiling technique in which coiling the ends at higher temperature is performed can reduce the property heterogeneities.
- May cause abnormal grain growth for a certain combination of finish rolling and coiling temperatures which causes defect in both hot and cold rolled sheets.

Thus, to obtain favourable texture and improved deep drawability and therefore, to increase annealed grain size, it is required that the rate of nucleation of recrystallized grains is reduced, which can be done by lowering of recrystallization temperature. This can be achieved by (i) reducing the carbon content, alloying elements and impurity elements and (ii) increasing the stored energy of deformation by higher cold deformation, adjusting the composition or hot rolling parameters to obtain desired distribution of hot band precipitates (**Lebrun et. al, 1981**) .Control of the dissolved carbon and carbide contents is achieved by lowering the carbon content of the steel, overageing after continuous annealing, coarsening of the cementite particles and reducing the rate of redissolution of the carbides during annealing. High temperature coiling also promotes coarsening of the carbides present.

Thus, contrary to batch annealing high temperature coiling improves texture and drawability for EDD steel in continuous annealing (Figure 4). In batch and continuous both, a temper rolling is recommended after annealing to remove yield point elongation and thus to avoid stretcher strains in the final product.

4. Recrystallization texture and formability for EDD steel

4.1 Heating rate effect

While the Al and N are kept in solution prior to annealing, the microstructure varies with heating rate and mechanical properties and r_m-value are strongly dependent on heating rate during annealing as represented by **Figure 5 (Beranger et al., 1996)**. Batch annealing involves placing sheet steel coils in a gas fired furnace with a controlled atmosphere. Batch annealing cycles normally involve slow heating up to about 700 °C. Slow heating after cold rolling is normally necessary to allow adequate time for the Al to diffuse, forming clusters or precipitates before recrystallization commences. Thus, low heating rate leads to the precipitation of AlN during recovery that helps generate strong {111} texture after recrystallization. The precipitation of AlN takes place at a lower temperature and this is followed by recrystallization of the steel at a higher temperature (Takahashi & Okamoto, 1974). The optimum heating rate up to the precipitation stage to obtain highest r_m-value was calculated by Takahashi and Okamoto (**Takahashi & Okamoto, 1974**): Log (PHR) = 18.3 + 2.7 log((Al) (N) (Mn)/R_{CR}), where PHR is the peak heating rate in Kh^{-1} corresponding to the peak in r_m-value, (Al), (N) and (Mn) are solute concentration in weight percent and and R_{CR} is the percentage reduction via cold working. The holding temperature is always below Ac1 that varies in the range of about 650-720°C (**Beranger et al., 1996**). The coils are then slowly cooled at 10°C/hour (Takahashi & Okamoto, 1974) and the process takes several days in batch annealing.

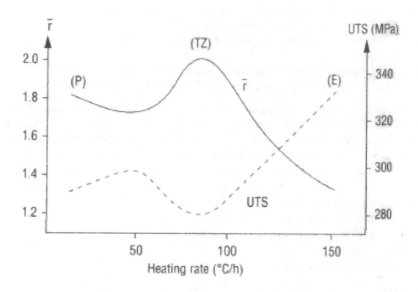

Fig. 5. Variation in r-bar (r_m) UTS values and microstructure with heating rate for batch annealed aluminium killed steel (Beranger et al., 1996).

Kozeschnik et al (**Kozeschnik et al., 1999**) calculated the logarithmic intensity ratio I_{111}/I_{100} at various annealing heating rates for the EDD steel. **Figures 6** (heating rate 50K/h) and **7** (heating rate -120K/h) show the effect of heating rate variation on the logarithmic intensity. Due to diminishing effective particle/recrystallization interaction time, the absolute value of the maximum intensity ratio is significantly decreased at higher heating rates. This effect is due to the shorter time available for aluminum nitride precipitation prior to the start of the recrystallization process. A higher supersaturation is needed in order to keep the two competing mechanisms balanced. Figure 8 shows the calculated variation of the I_{111}/I_{100} ratio with different heating rates. The qualitative comparison of the calculated logarithmic X-ray intensity ratios given in **Figure 8** and the experimental data for the r_m-value of Al-killed steel processed at low coiling temperature in Figure 9 (Hutchinson, 1984) shows good agreement.

Sarkar et al (**Sarkar et al., 2004**) also studied the effect of annealing heating rate on 70% cold rolled 0.05-C-0.19Mn-0.008 S-0.051Al EDD steel. In their work they selected heating rates of 50 and 70ºC/h to the intermediate annealing temperatures of 550 and 600ºC, and 15 and 30ºC.h from intermediate annealing temperature to the final annealing temperature of 700ºC. The soaking time at the intermediate and final annealing temperatures was half an hour. Tensile test and formability test inferred that the Annealing cycle that consisted of heating rate, 50ºC/h to 600ºC intermediate annealing temperature and 30ºC/h up to 700ºC resulted in the best combination of mechanical and formability properties with enhanced r_m of 1.92 and plane strain forming limit of 35%. The attractive combination of properties with high formability was attributed to the huge number of small spherical carbide precipitates that resulted in pure ferrite devoid of solute carbon.

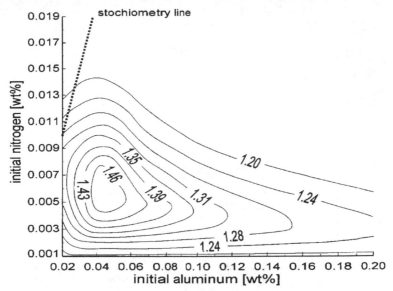

Fig. 6. Calculated logarithm of the intensity ratio (111)/(100) for a coiling temperature of 550 °C and a heating rate of 50 K/h as a function of initial amount of aluminum and nitrogen. [Kozeschnik et al., 1999]

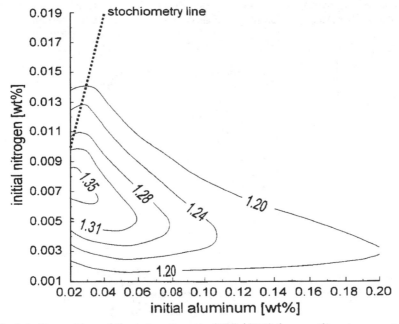

Fig. 7. Calculated logarithm of the intensity ratio (111)/(100) for a coiling temperature of 550 °C and a heating rate of 120 K/h as a function of initial amount of aluminum and nitrogen. [Kozeschnik et al., 1999]

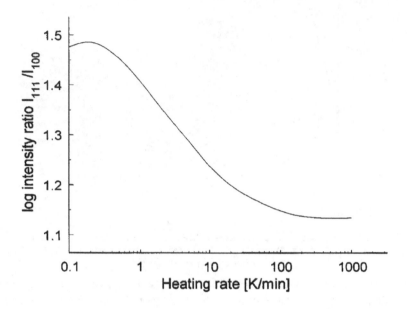

Fig. 8. Calculated logarithm of the intensity ratio (111)/(100) as a function of the heating rate during annealing for Al_{solute} = 0.03 wt pct and N_{solute} =0.005 wt pct. **(Kozeschnik et al., 1999).**

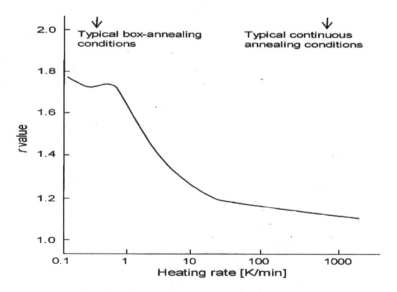

Fig. 9. Variation of mean plastic strain ratio, r_m with heating rate for EDD steel with low coiling Temperature **(Hutchinson,1984).**

4.2 Manganese and sulfur effect

Figure 10 (a, b) (**Ray & Jonas, 1990**) shows the comparison between the ODFs obtained from high Mn-high S (0.31%Mn, 0.018%S) and low Mn-low S (020%Mn, 0008%S) cold rolled batch annealed EDD steels, respectively. With the advances of steel making technology with desulphurization, the required Mn content to tie up with sulpher also reduced. The above technology in association with somewhat higher cold reduction, helped develop stronger γ-fiber (Fig 10b) that led to r_m, 1.8. While the r_m value obtained for the high Mn-high S steel (Fig 10a) was 1.5.

In the case of Al-killed steel, the optimized sheet properties are attained as a result of interaction between two processes: aluminium nitride precipitation and recrystallization (**Schulz, 1949**). Apart from pancaking of grains, the number distribution and morphology of carbides also play a role in formability. Smaller, spherical, and larger in number of carbides give good formability, because in this way the ferrite contains less C (ferrite is purer), which helps formability. Therefore, in order to improve the property of deep drawing sheet it is important to control every processing step effectively.

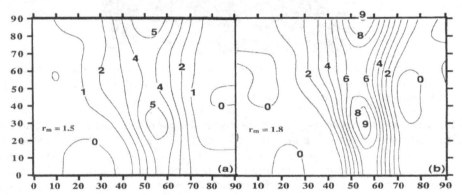

Fig. 10. φ2 = 45° sections of CRBA EDD steels showing (a) the ODF for the high Mn high S (0.31%Mn, 3 0.018%S) steel and (b) the ODF for low Mn, low S steel (0.20%Mn, 0.008%S) (Ray & Jonas, 4 1990).

4.3 Chromium effect

Mendoza et al (**Mendoza et al., 2004**) developed a chromium stabilized EDD steel (C: 0.02, Mn: 0.2, Al: 0.04, Cr: 0.35) by electric arc furnace, vacuum degassing, ladle treatment and continuous casting route. The steel sheet after cold reduced by ~85% was isothermally annealed under a protective argon atmosphere and was annealed at 700 °C from 1 to 300 s. The heating and cooling rate of annealed sheet specimens were ∞10 and ∞80 °C/s, respectively.

Figure 11 shows the SEM-micrograph of the as-cold rolled specimen. As may be observed, the ferrite grains are flattened, and inside the grains some shear bands can be observed. These in-grain shear bands corresponded to the narrow regions of intense shear that carry large strains during deformation and appear to become the major deformation mode (**Park et al.,2000**). In some interstitial free-steels also, it was observed that in-grain shear bands were inclined at angles of 30°-35° to the rolling plane (Barrett & Jonas, 1997). For instance, Barnett and Kestens (**Barnett et al., 1999**) reported that the increasing density and severity of these in-grain shear bands lead to a bulk recrystallization texture dominated by {1 1 1}<1 1 2> near the normal direction–rolling direction (ND–RD) for low carbon, ultra low carbon and interstitial free steels. In the present chromium stabilized EDD steel, the shear bands were inclined ∽32° with respect to the rolling plane and were almost parallel to each other within the individual grains.

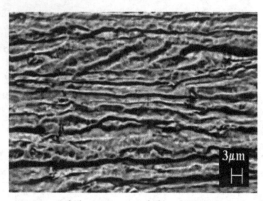

Fig. 11. Hot band microstructure of chromium stabilized EDD in hot rolled plates with flattened ferrite grains closely parallel to the sheet plane and inside the grains, some shear bands are observed (**Mendoza et al., 2004**).

Figure 12 shows {1 0 0} pole figures for the steel in the hot rolled, cold rolled and annealed conditions. It may be observed that in the cold rolled sheet, a {5 5 4}<2 2 5> component was noticed near {1 1 1}<1 1 2> texture. While, in the annealed sheet an appreciably strong texture than in the as cold rolled condition was observed from {5 5 4}<2 2 5> to {2 1 1}<0 1 1>. The microstructure obtained in the annealed condition was partially recrystallized, which was attributed to the presence of chromium-carbides precipitates that helped retard the recrystallization rate of the low carbon Al-killed/Cr-stabilized steel and which eventually was instrumental in the formation of {1 1 1}<1 1 2> textures.

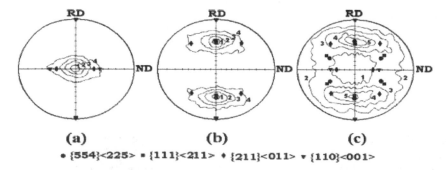

Fig. 12. {1 0 0} pole figures for the chromium stabilized EDD steel in (a) hot rolled, (b) cold rolled and (c) annealed conditions (**Mendoza et al., 2004**).

The mechanical properties of the annealed chromium stabilized EDD sheet are depicted in **Figure 13**. The use of chromium instead of niobium or titanium to stabilize the low carbon steel was effective in slowing down the recrystallization rate, thus, enhancing the formation of {1 1 1}<1 1 2> textures and achieving $\bar{r}$ values (r_m- values) >2 (Mendoza et al., 2004).

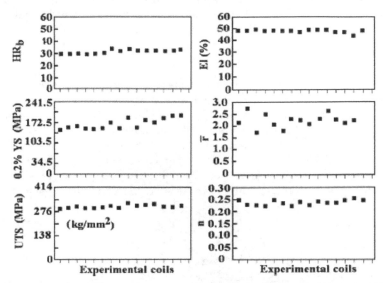

Fig. 13. Mechanical properties of annealed coils. (a) Hardness Rockwell b, (HRb), (b) 0.2% of yield strength, (YS), (c) tensile strength, (TS), (d) percent of elongation, (El), (e) the average plastic anisotropy value, ($\bar{r}$) and (f) the strengthening hardening exponent, n (**Mendoza et al., 2004**).

4.4 Phosphorous effect

It was observed by researchers that rimmed steel with P addition and by annealing the cold rolled sheet in decarburization furnace improved formability and thus studies were made on EDD steels to examine the interactive effect of P and AlN precipitation on formability of the steel sheets (Beranger., et al).

4.4.1 Heating rate effect on phosphorous added EDD steel

Rephosphorized Al-killed (EDD-P) steels show the similar dependence on heating rate of grains to conventional Al-killed steels (**Ono et al., 1982**). However, the smaller grain size, lower grain elongation ratio, lower r_m-value, faster recrystallizatio rate observed in the EDD-P steel were all attributed to the effect of phosphorous on the precipitation site of AlN. If phosphorus weakens the retardation effect of AlN on recrystallization, the restriction on nucleation of less favourable orientations, other than the {111} nucleus by AlN, which is usually observed in EDD steels will be relieved, resulting in the reduction of (111) intensity and an increase in the {100} intensity. In EDD-P steel {111}<112> orientation is observed unlike {111}<110> in the conventional EDD, which was attributed to the modification of cold rolled microstructure by phosphorous.

Figure 14 (Ono et al., 1982) shows the change in X-ray integrated intensity during heating. From the early stage to the half way of recrystallization, a rapid increase in the {222} intensity in association with a rapid decrease in the {200} and {110} intensities were observed in the EDD-P steel (Steel−1: C: 0.05, Mn-0.24, P:0.069) as well as in the conventional one (Steel-4: C:0.05, Mn:0.26, P:0.016). However, at the end of recrystallization, the EDD-P steel showed a lower {222} intensity and a higher {200} intensity than the conventional one. There was a marginal difference in the {110} intensity between them.

4.4.2 Manganese effect on phosphorous added EDD steel

An optimum combination of phosphorus and manganese content in steel can render a strong {111} texture through a simulated batch annealing cycle. As per Hu and Goodman (**Hu & Goodman, 1970**) and Hughes and Page (**Hughes & 1971**) the interaction between manganese and carbon would affect the recrystallization kinetics and thus, recrystallization textures. Further, Matsudo et al. (**Matsudo et al, 1984**) reported the detrimental effect of the Mn-C interaction on the drawability. On the other hand, it is widely known that steels containing phosphorus are likely to show a banded structure with segregated P-bearing ferrite, suggesting phosphorus and carbon in steel can act as repulsive elements to each other.

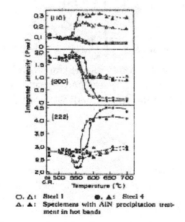

O, △: Steel 1   ●, ▲: Steel 4
△, ▲: Specimens with AlN precipitation treatment in hot bands

Fig. 14. Change in integrated intensity of Steels 1 and 4 during continuous heating at a peak heating rate 50°C/hr. (**Ono et al., 1982**).

Hu (**Hu, 1977**) reported that the drawability of low-carbon steel containing 0.067 pct P diminished more gradually than that of P-free steel with increasing manganese content up to 0.3 pct.

However, the work conducted by Chung et al (**Chung et al., 1987**) to examine the effect of manganese on the development of {111} recrystallization textures of P-containing low-carbon Al-killed steel sheet demonstrated that the phosphorous addition could modify the Mn-C interaction favourably for achieving improved drawability in EDD steels. **Figure 15** shows the (200) pole figures for the 0.1 pct P and P-free steels annealed at 973 K for 3 hours. The P-containing steels manifested {554} (225)-type texture and the texture was extremely strong in the 0.78 pct Mn-0.1 pct P steel. On the other hand, in the P-free steels the texture was not as strong at 0.5 pct Mn as it was at 0.06 pct Mn.

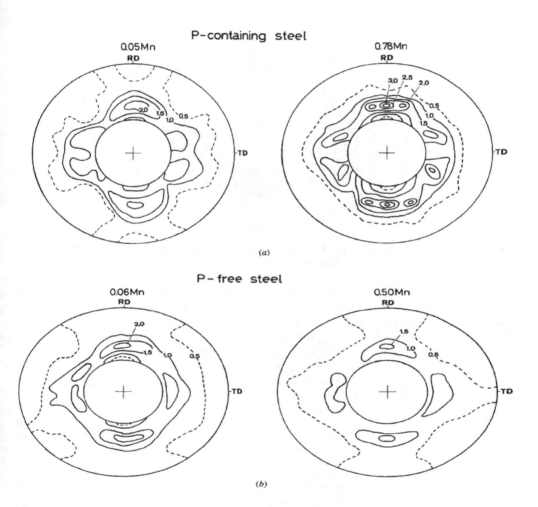

Fig. 15. (200) pole figures for (a) 0.1 pct P steels normalized, cold rolled, and annealed at 973 K for 3 h and (b) P-free steels normalized, cold rolled, and annealed at 973 K for 3 h (**Chung et al., 1987**).

To confirm the effectiveness of phosphorus addition to a high manganese steel in order to get a high r_m-value, the 0.2 pct P-1.2 pct Mn steel was employed. The recrystallization texture, as shown in **Figure 16**, had strong {554}(225) components which were favorable for a high r_m-value. The r_m-value and other mechanical properties are shown in **Table 1 (Chung et al., 1987)**. The high manganese P-containing steel had a tensile strength of 430 MPa and an r_m-value of 2.0.

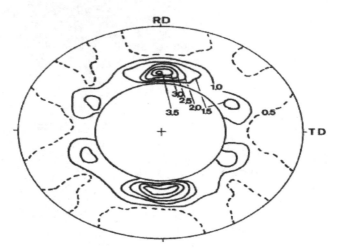

Fig. 16. (200) pole figure of 0.2 pct P-1.2 pct Mn steel preaged at 843 K for 24 h, cold rolled 75 pct, and annealed at 973 K for 4 h (**Chung et al., 1987**).

The micrographs of the P-free steel and the P-containing steel hot rolled and aged are shown in **Figures 17** and **18**, respectively. In the P-free steels, the precipitates that were identified as AlN dispersed within the grain at 0.06 pct Mn, whiles the precipitates were barely noticed at 0.5 and 1.0 pct Mn. On the other hand, in the P-containing steels, precipitates which might be $(FeMn)_3C$ and/or AlN were dispersed uniformly in the ferrite matrix only at the 1.5 pct Mn level. From the results, it must be noted that the fine precipitates of nitrogen and/or carbon disperse uniformly with decreasing manganese in the P-free steels, and vice versa with increasing manganese in the P-containing steels.

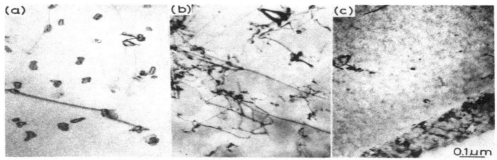

Fig. 17. Transmission electron micrographs of P-free Al-killed steels aged at 833 K for 1000 h after hot rolling. (a) 0.06 pct Mn, (b) 0.5 pct Mn, and (c) 1.0 pct Mn (**Chung et al., 1987**).

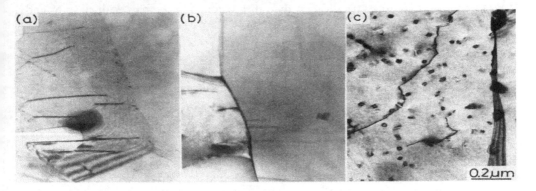

Fig. 18. Transmission electron micrographs of 0.2 pct P steels aged at 843 K for 24 h after hot rolling. (a) 0.1 pct Mn, (b) 0.5 pct Mn, and (c) 1.5 pct Mn **(Chung et al., 1987)**.

Y.S. (MPa)	T.S. (MPa)	Uniform El. (Pct)	Total El. (Pct)	r_0	r_{45}	r_{90}	$\bar{r}$
262	434	24	33	1.60	2.09	2.25	2.01

Table 1. Mechanical properties of 0.2 pct P-1.2 pct Mn Al-Killed steel preaged at 8543 K, cold rolled 75 pct, and annealed at 973 K for 4 hrs.

Thus, a strong recrystallization texture with {554} (225) texture component and an uniform dispersion of $(FeMn)_3C$ in ferrite matrix can be developed through batch annealing of a P-containing EDD- steel if the steels contain high P and high-Mn .

5. On metallurgy of ultra-low carbon interstitial-free (IF) steels

Interstitial free steels are highly formable due to their low carbon and nitrogen content of ((wt%) < 0.003% and < 0.004%, respectively). The C and N are tied to Ti, Nb etc as precipitates. A limited excess of titanium or niobium relative to carbon, nitrogen and sulfur contents has a favourable influence on mechanical properties. During hot strip rolling, the level of interstitial elements, such as C and N, remnant from the steel making process, can be reduced by combining them with the stabilizing elements. The application of these steels is in the rear floor pan, front, rear door inners, etc.

5.1 Titanium stabilized IF steels

Titanium is very effective in scavenging nitrogen, sulphur, and carbon, readily forming TiN during casting and TiS during slab reheating . Subsequently, while the nitrogen and sulphur are scavenged, remaining Ti ties up with TiC during coiling. The minimum amount of titanium required for full stabilization, based on a stoichiometric approach, is (**Tither & Stuart, 1995**):

$$Ti_{stab} = 4C + 3.42N + 1.5S$$

It was proposed that excess Ti addition than that required to combine with all C, N, and S was beneficial to achieve high r values (**Gupta & Bhattacharya, 1990**). Excess Ti (Ti*) is given by :

$$Ti^* = Ti_{total} - (4C + 3.42N + 1.5S)$$

However, it is also commented that an excess Ti content can be linked to the incidence of surface streaking. The frequency of appearance of this type of defect can be minimized using Nb in combination with Ti. Titanium only intierstitial free steels are the least susceptible to compositional changes and process variation. This is attributed to the precipitation of Ti compounds at higher temperatures (TiN, TiS) that play less of a role in subsequent lower temperature processing (Krauss et al., 1991). Tsunoyama et al (**Tsunoyama et al, 1988**) considered three precipitation mechanisms suggested by various authors: a) TiS may provide a preferential site for TiC nucleation. Thus, lowering the S content can retard the precipitation of TiC b) with decreasing Ti content, precipitation of $Ti_4C_2S_2$ occurs in place of TiS and controls carbon stabilization. The stabilization of C by $Ti_4C_2S_2$ is preferred compared to TiC precipitation as $Ti_4C_2S_2$ precipitates are larger, remove solute carbon from the matrix earlier, and are more stable.

The precipitation mechanism of titanium in these steels is summarized in **Figure 19** (**Tither, 1994**). It is postulated that titanium nitride particles formed during slab casting acted as nucleation sites for the precipitation of TiS and $Ti_4C_2S_2$. The small amount of carbon remaining is precipitated as TiC. During reheating of the slab, solution of carbosulphide occurs, leaving only TiS and TiN. Cooling of the strip to the intercritical (austenite -ferrite transformation) temperature region during hot rolling transforms TiS to $Ti_4C_2S_2$ by the absorption of titanium and carbon.

5.2 Niobium stabilized IF steels

In Nb IF steels, Nb forms carbonitride precipitates. The aluminium addition during the killing process also reacts with N to form AlN. However, the favourable solubility product of AlN compared to NbCN, leads to preferential precipitation of AlN at a higher critical temperature. This reduces the N available for the precipitation as NbCN. The solubility products are dependent on bulk chemistry, temperature, and precipitate composition. The precipitates sequence in such steels can be considered as (**Holie, 2000**)

$$Al + N = AlN$$

$$Mn + S = MnS$$

$$Nb + C + N = NbC$$

The solute Nb on grain boundaries introduces site competition for the elements like phosphorous, which enables lower ductile to brittle transition temperatures, poorer elongation and r_m- values than Ti IF steels. The niobium steels have higher recrystallisation temperature (750 - 800°C) than titanium stabilized IF steels (**Baker et al., 2002**) and it was reported that the temperature of recrystallisation of Nb IF could be up to 50 K higher than that of the Ti IF steels. This in turn requires higher finish rolling and annealing temperatures. The higher recrystallization temperature is attributed to the low temperature formation of niobium carbides, producing fine particles that readily retard grain boundary movement during annealing.(**Tsunoyama, 1990**). The temperature can be reduced by lowering C content. In addition, recrystallisation temperature can be reduced to enhance drawability, using higher coiling temperature, so that precipitates are coarsened and subsequent pinning effect is reduced.

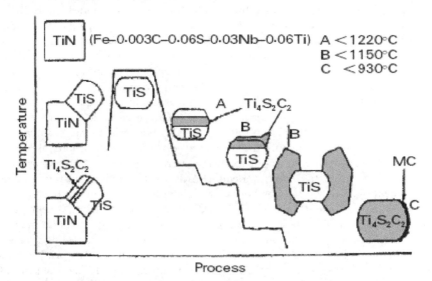

Fig. 19. Schematic mechanism of carbosulphide and carbide formation in titanium stabilized IF steels (**Tither et al., 1994**).

5.3 Titanium and niobium stabilized IF steels

Titanium in association with niobium added IF steels have the advantages of both Ti only and Nb only IF steels with reduced disadvantages. This combination provides the best combination of properties (**Tokunaga 1990**). The levels of Ti and Nb additions that are required to stabilize Ti +Nb IF steels are given by (**Holie, 2000; Tokunaga & Kato, 1990**)

$$(\%Nb)-(7.75(\%Ti)-3.42(\%N)-1.5(\%S))/4=0$$

If insufficient Ti is added such that the level of addition falls below that required for stabilization, TiN forms first and any S that is left behind after the formation of TiS is consumed as MnS. If the steel is understabilised by the insufficient addition of Ti and/or Nb then excess C exists in solid solution and a bake hardenable steel may be produced.

For IF steels, the increase in r_m-value with degree of cold work is continuous over the whole range employed in industrial production (**Beranger et al**).An increase in annealing temperature and/or holding time improves the deep drawability due to grain growth and recrystallization texture enhancement. Further, these grades have very high Ac1 temperatures, enabling continuous annealing at temperatures of ~850°C, or even higher. For such steels, the metal has no yield point elongation and no requirement of overaging treatments in continuous annealing, due to the absence of interstitials C and N.

5.4 Phosphorous effect on IF steels

One of the ways of increasing the strength of the IF steels is addition of solid solution strengtheners viz. P, Si and Mn to the steel (**Rege et al., 2000**). It has been found that P is the most potent and cost effective solid solution strengthener that increases strength without appreciably affecting the drawability of the steel (**Katoh et al., 1985; Tokunaga & Kato, 1990**). Mn, on the other hand, significantly deteriorates drawability and ductility (**Irie et al. 1981; Katoh et al., 1985; Tokunaga & Kato, 1990**), and Si adversely affects coating adhesion (**Nishimoto et al., 1982**). Thus, P is the preferred addition to increase the strength of IF-steel. However, it is found that phosphorus tends to segregate at the grain boundaries or precipitate out of the matrix during the recrystallization annealing. The phosphorus segregations reduce the cohesive strength of the grain boundaries, which leads to the secondary cold work embrittlement (CWE) and reduces the resistance of the steel to brittle fracture or more precisely, intergranular fracture (**El-Kashie et al. 2003; Rege & DeArdo, 1997**). The CWE is defined as the susceptibility of the sheet material to intergranular fracture during the secondary work of deeply drawn part or while in service. Cao et al (**Cao, 2005**) had reported that the segregation of phosphorus occurred when the P>0.07 wt pct in the IF steel. It is also found that the batch annealing leads to the higher phosphorus than the continuous annealing. The segregation behavior of phosphorus in the Ti and Ti+Nb IF steels was studied by Rege et al (**Rege et al., 2000**). It was found that the segregation of phosphorus to the ferrite grain boundaries occurred not only during the coiling stage of the thermo-mechanical processing, but also during the cold rolling and annealing process. The steels with higher phosphorus content showed higher ductile to brittle transition temperature, i.e, lower resistance to cold work embrittlement (CWE). The problem of CWE is more often encountered while annealing time is long, which enables P to segregate in the grain boundary (**Yasuhara, 1996**).

The CWE can be avoided by :

- controlling the chemical composition−i. e. by partial stabilization of carbon in IF steels or by addition of B/Nb - It is believed that P and C both segregate to the grain boundary and compete for the available sites, and C reduces the grain boundary segregation and embrittlement by P. Furthermore, C enhances the grain boundary cohesion and counteracts the embrittlement this way. B also plays the same role as that of C. Hence P-C or P-B site competitive process was expected to minimize the CWE phenomena. In Nb added steel it is believed that CWE is lower than that in Ti-stabilized steel that has been attributed to the partial stabilization of C in Nb-containing steel.
- grain boundary engineering−i. e. by grain boundary character distribution−by suitable annealing cycles low angle and low -CSL boundaries can be produced by avoiding the development of continuous random boundary network and which help reduce CWE.

6. Recrystallization texture and drawability of IF steels

Cold rolling plays an important role in the formation of favourable textures for deep drawing during annealing, however, has little effect on other properties. The variation of r-value with cold reduction in the three common IF steel types, Ti-stabilized, Nb-stabilized and Ti-Nb stabilized is shown in **Figure 20** (**Tokunaga & Yamada, 1985**). Ti -

Nb steels show the highest r-value for equivalent cold reduction. This was attributed to the fact that the precipitates formed during hot rolling and coiling were not large enough to compromise the r-value during annealing. A reduction of 90% produced the highest r-value in all the steels, however, these reductions were rarely achieved in practice, 80% being more common.

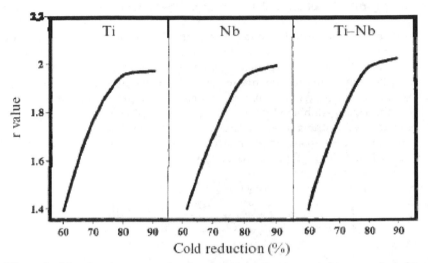

Fig. 20. Effect of cold reduction on r-value for titanium, titanium-niobium, and niobium IF steels (**Tokunaga & Yamada, 1985**).

Mendoza et al (**Mendoza, 2000**) studied the hot rolled precipitation behaviour of Ti-stabilized IF steels and mechanical properties and reported about improved r_m-value of 2-2.14. While Banerjee et al (**Banerjee et al., 2008**) vividly studied the precipitation of hot-band and annealed Ti stabilized IF steels to optimize hot and cold spot temperatures during annealing for improving drawability of the steels to >1.9. The hot bands of the as-received steels showed two types of texture development: (a) shear texture, manifested by {225}<554>, {110}<100>, and {113}<332> and (b) cold deformation type texture with a strong and incomplete α- fiber consisting of the components {001}<110>, {112}<110>, and {111}<110>, and the γ-fiber components, {111}<110> {111}<112>. The presence of a substantial amount of coarse equiaxed ferrite grains at the surface and finer grains at the central region was attributed to the outcome of finish rolling in the two-phase α+ γ region. On the other hand, the deformed grains at the surface and finer grains at the central region another hot band was the result of rolling in the single- phase α region. Figure 21 (Banerjee 21 et al., 2008) shows the φ2= 45° orientation distribution functions for various combinations of cold reduction and batch annealing temperatures to obtain improved r_m-value in an optimized processing condition. The poor texture and r_m-values at the annealing temperature of 660 °C were associated with the precipitation of fine precipitation of FeTiP-type within the grains and at the grain boundaries. This study had shown that for the given chemical composition of the Ti-IF steel, the optimized condition for cold spot temperature in batch-annealing cycle was 680 °C, preceded by 80 pct cold reduction, which resulted in an r_m-value of as high as 2.29.

The influence of the texture development in Ti-added (0.03, 0.05 and 0.07 wt%) IF on r_m-value was investigated by Kim et al (**Kim et al., 2005**). It was intended to determine the optimized Ti content for the promotion of deep drawability in the IF steels. For the IF steel with the composition of 0.0025C, 0.070Mn, 0.002N and 0.007S, the optimum Ti content was found to be 0.05wt%.

Juntunen et al (**Juntunen et al., 2001**) investigated the continuous annealing parameters in laboratory scale on drawability of Ti+Nb stabilized IF and IF-HS steels and it was reported that r_m-value could be enhanced by about 13% simply by adjusting the annealing conditions. The annealing cycle with maximum studied temperature produced the sharpest γ-fiber and highest r_m-value. While, Ruiz-Aparicio et al (**Ruiz-Aparicio, et al., 2001**) studied the evolution of the transformation texture in two $Ti_4C_2S_2$-stabilized interstitial-free (IF) steels (Ti and Ti/Nb) as a function of different thermomechanical processing parameters. Analysis showed that the $Ti_4C_2S_2$-stabilized steels stabilized were not very sensitive to the processing conditions employed in the study. The study also revealed that, under conditions of large deformations and coarse austenite grain sizes, the main components of the transformation textures are the beneficial {111}||ND orientations.

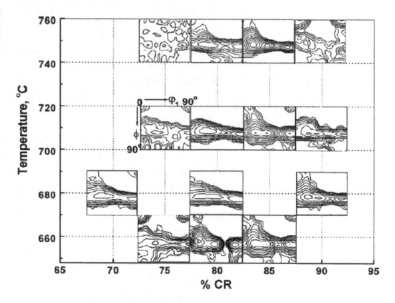

Fig. 21. ODFs of φ2 = 45° section at various processing conditions for a particular batch annealing temperature at various cold reductions. Maximum intensity for **660** °C: 75 pct-3.53, 80 pct -8.43, and 85 pct-6.43; for **680°C** (hot spot): 70 pct-9.68, 80 pct-12.13, and 90 pct-10.38; for **710** °C: 75 pct-2.95, 80 pct-8.09, 85 pct-6.89, and 90 pct-4.13; and for 750 °C: 75 pct-2.0, 80 pct-10.77, 85 pct-9.46, 90 pct-3.32 (**Banerjee et al., 2008**).

The effects of electric field annealing on the development of recrystallization texture and microstructure in a Ti+Nb stabilized cold rolled IF steel sheet were studied by He et al (**He et al., 2003**) means of ODF analysis and optical microscopy to assess the drawability response of the steel. Specimens of size 50 mm X 20 mm were cut from the sheet with the longitudinal direction parallel to the rolling direction. They were then subjected to isothermal annealing at different temperatures ranging from 650 to 800 °C for 15 min, respectively with or without a DC electric field of 200 V/mm. The annealing treatments were done in a nitrogen atmosphere and at a heating rate of 5 °C/min to the chosen peak temperatures. The external electric field was applied by placing the specimens (positive electrode) in the middle of two parallel stainless steel sheets (negative electrode) that were 2 cm apart. The experimental arrangement is shown in **Figure 22**.

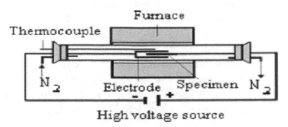

Fig. 22. Experimental arrangement for annealing with electric field (**He et al., 2003**).

Heating rate annealing experiments were carried out by Muljono et al (**Muljono et al., 2001**) to study the effect of heating rate on the recrystallization kinetics, grain size and texture of steels with a range of carbon levels (0.003-0.05% C). The steels were cold-rolled to 70% reduction and subsequently annealed at heating rates from 50 to 1000°C/sec to temperatures in the range 600 to 900°C. **Figure 23** shows that, for the 0.02 and 0.003% C steels, the {111}||ND texture increases in strength with increased heating rates up to 200°C/sec and maintains a plateau thereafter. Both steels exhibit similar trends and only the relative strength of the γ-fibre differs. The grain size and texture results are in general agreement with work by Hutchinson and Ushioda (**Hutchinson & Ushioda, 1984**). The {111}||ND components of the recrystallization texture increased at rates up to 200°C/sec due to enhanced nucleation at grain boundary sites.

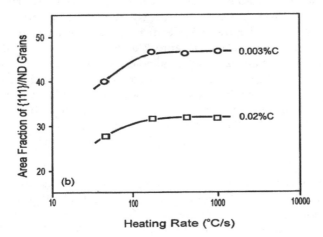

Fig. 23. Effect of heating rate on strength of recrystallization texture, given as the area fraction of grains within 15° of{111}||ND (**Muljono et al., 2001**).

During electric field annealing, specimens work as anode and the applied electric field reduces the lattice defect energy by lowering the shield effect that would decrease the driving force for recrystallization (**Cao et al.,1990; Conrad, 1989; Wang, 2000**). Thus, although the application of electric field generally reduces the driving force for nucleation and grain growth, the nuclei with random orientations are much more restricted than that of the γ-nuclei. Consequently, the electric field annealing yields a high nucleation rate for the γ-nuclei that lead to a relatively strong γ-texture after complete recrystallization. From Figures 24 and 26 (**He et al, 2003**) it can be noted that both kinds of specimens annealed with and without application of electric field, exhibited a similar tendency in the development of recrystallization textures, i.e. the α-fiber was weakened and the γ-fiber was strengthened with increasing annealing temperature. In addition the Figures 4 & 6 also depict that the application of electric field (200 V/mm) during annealing may promote the development of the γ-fiber (ND | | <111>) recrystallization texture of the cold-rolled IF steel sheet, which is beneficial to the deep-drawability. While, Figures 25 and 27 (**He et al, 2003**) indicate that the recrystallization was noticeably retarded intensively by electric field annealing under the investigated conditions.

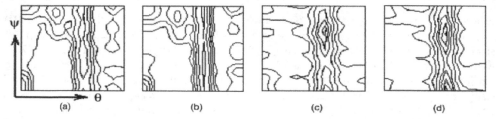

(a)   (b)   (c)   (d)

Fig. 24. φ= 45º sections of the ODFs (levels: 1; 2; 3; . . .) for the specimens annealed without an electric field at (a) 650 ºC, (b) 700 ºC, (c) 750 ºC and (d) 800 ºC (**He et al, 2003**).

In an innovative finding, Jeong communicated (**Jeong, 2000**) that r value was markedly improved by reducing the carbon content from 0.0035 (Steel B) to 0.0009 pct. (Steel A) in Ti stabilized IF steels that were treated with 0.25 pct Si, 1.25 pct Mn, and 0.09 pct P to attain tensile strength of 400 MPa (**Figure 28a**). The difference in r_m-value between two steels is 0.2 to 0.3 at all annealing temperatures (**Figure 28b**). Steel A containing 0.0009 pct carbon showed a high r_m-value of about 1.6 while annealed at 800 ºC to 860 ºC, corresponding to deep drawing quality (DDQ). The r_m-value increased to 1.85 for extra deep drawing quality (EDDQ) grade with increasing annealing temperature to 890 ºC. It was thus a remarkable finding as the steel was high strength steel with tensile strengths of 400 MPa or higher. This result indicated that while the carbon content decreases below 0.001 pct, superior formability of EDDQ grade could be achieved in high strength steel with tensile strength of 400 MPa or higher.

The highest $\bar{r}$ (r_m-values) in steels A and B were obtained in the specimens annealed at 890ºC while the coarsening of the ferrite grain was remarkable (**Figure 28c,d**). In order to find out the reason for the effect of carbon on the r_m-value, (200) pole figures for the annealed sheets were measured. The comparison of (200) pole figures of steels A and B showed that with the decrease in the carbon content, {554}{225} near the ND//(111} texture became stronger that was responsible for the improvement of r_m- value (Jeong, 2000).

Song et al (**Song et al., 2010**) worked on phosphorous segregation and phosphide precipitation on grain boundaries in association with drawability of rephosphorised IF steel .The cold rolled steel was annealed at 810ºC for 90 to 600 sec in a protected environment. The **Table 2 (Song et al., 2010)** illustrates yield strength, tensile strength and r_m-value reduce while n (work hardening index) value increases with the annealing time from 180 to 600 s. The phosphorus concentration at grain boundary is shown in **Figure 29 (Song et al., 2010)**. It was observed that as annealing time increased, the phosphorus concentration increased from 0.22 to 0.8 wt pct. The phosphorus concentration at the grain boundary is 20 times higher than that in the matrix for the sample annealed for 600 s, which greatly reduces the steel strength.

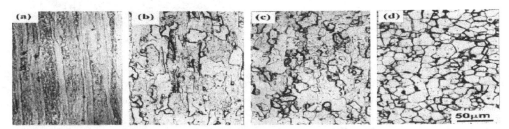

Fig. 25. Microstructures of the specimens annealed without an electric field at (a) 650 ºC, (b) 700 ºC, (c) 750 ºC and (d) 800 ºC (**He et al, 2003**)

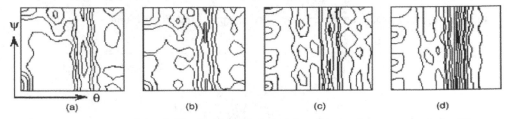

Fig. 26. $\phi= 45º$ sections of the ODFs (levels: 1; 2; 3; . . .) for the specimens annealed with an electric field at (a) 650 ºC, (b) 700 ºC, (c) 750 ºC and (d) 800 ºC (**He et al, 2003**).

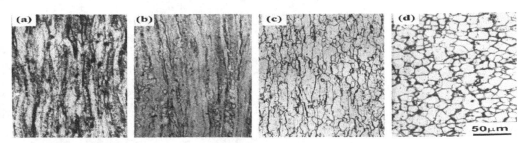

Fig. 27. Microstructures of the specimens annealed with an electric field at (a) 650 ºC, (b) 700 ºC, (c) 750 ºC and (d) 800 ºC (**He et al, 2003**).

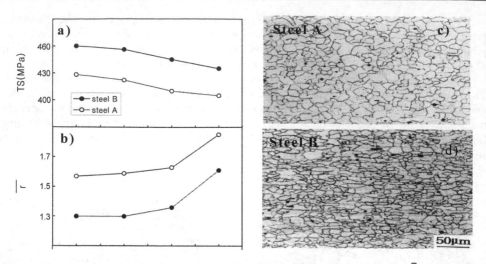

Fig. 28. Effects of carbon and annealing temperature on (a) tensile strength (b) $\bar{r}$ (r_m-value) and (c) & (d) ferrite grain sizes after annealing at 890°C for Steels A and B (**Jeong, 2000**).

Annealing time/s	Yield strength/MPa	Tensile strength/MPa	r	n
180	158.1	354.9	2.098	0.2376
360	155.4	346.8	1.939	0.2936
600	141.1	322.7	1.688	0.3046

Table 2. Mechanical properties of the annealed samples for different times

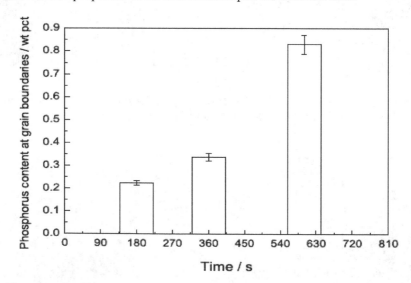

Fig. 29. Phosphorus content at grain boundaries in the rephosphorised IF steels annealed for different time 5 (**Song, et al., 2010**)

Kang et al (Kang et al., 2007) studied on the effects of aluminium on interstitial free high strength steel for the improvement of drawability (Figure 30). 78%cold rolled samples were annealed in an infrared-ray heating furnace. The annealing cycle consisted of heating the specimens to 830°C at a constant heating rate of 7°C/s and held at temperature for 30 s and then cooled to RT. Aluminum content more than 0.10 wt% improved the formability of the IF-HS. Texture analyses showed that the {111}||ND fiber (γ-fiber) was intensified, and <110>||RD (α-fiber) was weakened, with the increase of aluminum content. Recrystallization was completed earlier in the steel with the high aluminum content and the grain size of the annealed sample was larger than the steel containing lower aluminium. It was confirmed thorough the SANS analysis that the size of the precipitates in the sample with higher aluminum content was larger and their number was much fewer than in the sample with lower aluminum content. It appears that the high aluminum content in IF-HS containing Mn, P, Ti and Nb improved the scavenging effect of Ti or Nb and thus purified the iron matrix.

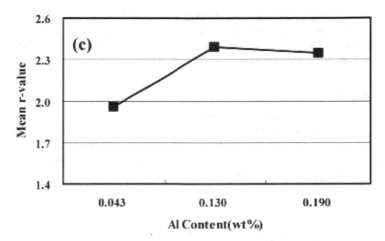

Fig. 30. Effect of aluminum content on mean r-value of the interstitial free high strength steel containing Mn, P, Ti and Nb (Kang et al., 2007) (Kang et al., 2007)

7. Metallurgy of bake hardening steels

Bake hardening is a diffusion controlled process involving the migration of solute carbon atoms within the iron lattice. The diffusion of these atoms is affected by heat treatment time and temperature and the amount of solute present in the steel. Factors such as grain size and dislocation density may also have an influence.

7.1 Mechanism of bake hardening

The yield strength increase from bake hardening is accompanied by the return of the yield point and yield point elongation; there may also be a slight increase in tensile strength and decrease in elongation. To attain such increased strength, the following criteria must be met: (Matlock et al., 1998)

i.   mobile dislocations must be present in the steel.
ii.   there must be sufficient concentration of solute in the steel to pin these dislocations.
iii.   the solute must be mobile at the aging temperature.
iv.   dislocation recovery must be sufficiently slow to prevent softening.

The driving force for pinning is a reduction in lattice energy. Both impurity atoms and dislocations induce lattice strains in the iron matrix and these strains can be relaxed if the interstitial atoms diffuse to the vicinity of dislocations (**Mizui et al., 1990**).Bake hardening steel is amply used by the automotive industry for the outer panel of cars. This steel grade is characterized by low yield strength prior to final manufacturing process and a remarkably enhanced yield strength of the finished product in association with excellent deep drawability. The increase in yield strength during bake hardening of steel occurs due to the blocking of otherwise mobile dislocations by forming Cottrell atmosphere of solute carbon or iron-carbide precipitates during the baking operation after painting at 170ºC for about 20 minutes (**Kozeschnik et al., 1999**).

There are several types of bake hardening steel in production today. Individual grades are determined by the processing technology available and the properties required in the final product. These steels can be loosely grouped into two main categories, bake hardening EDD (aluminium killed) steels and bake hardening IF steels.

7.2 Bake hardening EDD steels

The EDD steel grades contain carbon content of the level~ 0.01%. If the carbon is allowed to remain in solution, room temperature aging occurs. Thus, carbon levels must be controlled by suitable annealing practices. These steels can exhibit problems during galvannealing, however, their main disadvantage is poor formability due to the presence of relatively high carbon contents (**Baker et al., 2002**). Careful chemistry control during steelmaking is therefore crucial to ensure suitable amounts of carbon remain in solution in the final product.

During slow cooling of batch annealed EDD grade steels almost complete precipitation of the carbon occurs and thus the remaining solute carbon is insufficient to cause bake hardening. To obtain bake hardening effect in batch annealed EDD, very low carbon grades of C~ 0.01% with some elements, like phosphorous, that increase carbon concentration in solution are employed. (**Beranger et al.**).

While in the case of continuous annealing, the carbon level in solid solution at the end of annealing is quite high and the solute carbon can produce strengthening (70-80 MPa) by bake hardening (steel book). Care must be taken, however, with overaging practice during annealing, to ensure an appropriate amount of carbon (15-25 ppm) is left in solution in the final product.

7.3 Bake hardening IF steels

IF steels contain very low amounts of total carbon (~0.004%) (**Baker et al., 2002**).. In these steels all the interstitial elements are removed from solution by addition of carbide and nitride formers such as aluminium, titanium, niobium. These steels do not exhibit bake hardening as they have no interstitial elements in solution. However, the chemistry and processing of these steels, can be adjusted to leave 15 -25 ppm carbon in solution, to render bake hardening effect for increasing strength by 30 - 60 MPa . There are several types of IF steels and in the following sections Ti-only, Nb-only and Ti+Nb steels will be discussed for bake hardening.

7.3.1 Titanium stabilized interstitial free bake hardening steels

This mechanism of precipitation applies only when some titanium remains in solution: in just stabilized or understabilized chemistries, formation of TiC and $Ti_4C_2S_2$ is inhibited because of the low titanium content (Baker et al., 2002). This successive precipitation of carbides and carbosulphides makes it difficult to control the amount of titanium available for carbon stabilization and thus the solute carbon content. However, a bake hardening product can be produced from a titanium chemistry by inhibiting or avoiding TiC formation, so that the total carbon content remains in solution and is available for bake hardening.

Work by Tanioku et al. (Tanioku et al., 1991) and Kojima et al. (Kojima et al., 1993) showed that this can be achieved by controlling total carbon at 15 -25 ppm and titanium at ~ 0.01%. The manganese content must be kept low (~ 0.3%) to prevent the formation of MnS in preference to TiS and the slab reheat temperature must be high (~ 1200°C) to prevent the precipitation of $Ti_4C_2S_2$. Another method for producing titanium IF bake hardening steels, discussed by Tsunoyama et al. (Tsunoyama et al, 1998) relied on reducing the sulphur level to minimize formation of TiS, as the role of TiS as a heterogeneous nucleation site for the precipitation of TiC can lead to the reduction in solute carbon and hence bake hardening response. Kawasaki et al. (Kawasaki et al., 1991) used a philosophy, involving a reduced sulphur level (0.005%) in association with an increase in manganese to 1.0% and thus, the formation of $Ti_4C_2S_2$ was suppressed, leaving carbon in solution and producing a bake hardening steel.

7.3.2 Niobium stabilized interstitial free bake hardening steels

Niobium is a strong carbide former that can stabilize carbon as NbC when added according to the stoichiometric ratio:(%Nb)=7.75(%C) (Baker et al., 2002).).

Nitrogen is stabilized by the addition of aluminium to form AlN. This is more stable and thus forms at higher temperatures than Nb(C,N), so it can be assumed that all niobium is available for carbide formation. The relative simplicity of carbon stabilization in these steels makes them ideal for the production of bake hardening grades. Control of solute carbon in the niobium bearing bake hardening steels can be achieved in two ways. First, insufficient niobium can be added fully to stabilize the carbon, leaving 15 - 25 ppm in solution after steelmaking. This methodology requires tight chemistry control during steelmaking and, because of the presence of solute carbon throughout subsequent processing, the rm-value can suffer.

The second method requires the full stabilization of carbon during steelmaking. Solute carbon is then liberated by solution of NbC during annealing. By annealing at high temperatures (800-850°C) and cooling at 420 Ks^{-1},15 -25 ppm carbon can be retained in solution (Irie et al., 1982). Since the carbon is fully stabilized until the end of annealing, r_m-

values in steels of this type are comparable with those of traditional IF grades. High temperature annealing of this kind can, however, lead to shape problems in the strip such as heat buckling. Some continuous annealing lines can operate at these high temperatures, but they are beyond the limits of conventional hot dip galvanizing lines. Steelmakers must therefore assess their own production capabilities before deciding on a suitable processing route for the niobium IF based bake hardening steels.

By reducing the sulphur level and increasing manganese, the formation of TiS can be suppressed, leaving all titanium available for nitrogen stabilization. The precipitation of $Ti_4C_2S_2$ can also be suppressed in this way, so carbon is controlled by niobium alone. As with the Nb only compositions, carbon content can be controlled either by understoichiometric addition of niobium, or by high temperature annealing and controlled cooling, depending on the capabilities of individual steelmakers. These steels have been widely researched and are the choice of many manufacturers.

7.3.3 Titanium and niobium stabilized interstitial free bake hardening steels

By reducing the sulphur level and increasing manganese, the formation of TiS can be suppressed, leaving all titanium available for nitrogen stabilization (**Baker et al., 2002**).). The precipitation of $Ti_4C_2S_2$ can also be suppressed in this way, so carbon is controlled by niobium alone. As with the Nb only compositions, carbon content can be controlled either by understoichiometric addition of niobium, or by high temperature annealing and controlled cooling, depending on the capabilities of individual steelmakers. These steels have been widely researched and are the choice of many manufacturers.

8. Recrystallization texture and drawability of bake hardening steels

Kitamura et al. (**Kitamura et al., 1994**) had presented a completely different methodology for the production of titanium based interstitial free bake hardening steels. The theory suggests that as the steel sheet absorbs carbon by annealing in a carburizing atmosphere, the Ti/C ratio decreases, eventually resulting in some solute carbon in the matrix. A bake hardening response of 20 -50 MPa was achieved in this way without compromising r value (**Figure 31**) (**Kitamura et al., 1994**). Through this process, the reduction in r-value due to solute carbon in solution for interstitial free bake hardening steels can be eliminated as the steel remains fully stabilized by titanium and the excess carbon is introduced by carburizing atmosphere during annealing.

Xiaojun and Xianjin (**Xiaojun & Xianjin, 1995**) developed a new technology (the details were not mentioned) to improve the drawability of Ti+ Nb stabilized interstitial free high strength bake hardened steel. The r_m-value of the experimental sheet treated by the new technology is as high as 2.67, and this is the highest r_m-value published so far for phosphorus-added high strength and deep drawing sheet steels with increased strength due to bake hardening. Compared to conventional technology (**Figure 32a**), the new technology annealing rolling texture (**Figure 32b**) exhibited strong {111} components and weak {100}. The crystal orientations corresponding to the peak values of orientation concentrations of the texture were found to be changed from conventional (111)(112) orientations to (111)(011) orientations for the new technology (Figure 33).

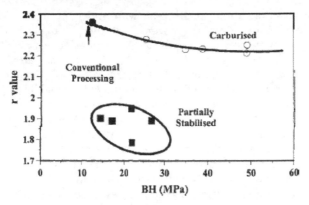

Fig. 31. Effect of processing on relationship between bake hardenability and r value for Ti-stabilized IF steel (**Kitamura et al., 1994**).

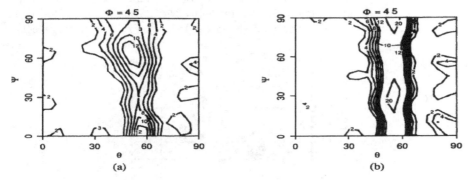

Fig. 32. φ=45° sections of ODFs of annealing textures obtained by two technologies respectively: (a) conventional processing and (b) new processing (**Xiaojun & Xianjin, 1995**).

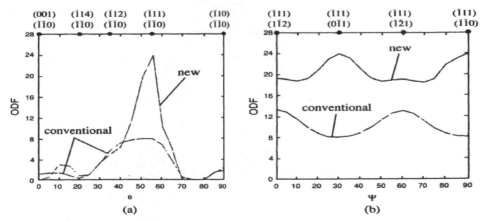

Fig. 33. Comparison of annealing textures obtained by two technologies respectively. (a) α-fiber axis textures (b) γ-fiber axis textures (**Xiaojun & Xianjin, 1995**).

Bake hardenability and drawability of IF steels with under- to over-stoichiometric atomic ratios of Ti/N and Nb/C were studied by Storojeva et al (http://www.cbmm.com.br/portug/sources/techlib/report/novos/pdfs/stabiliza.pdf). They observed that the r_m-value increased with a larger ferrite grain **(http://www.cbmm.com.br/portug/sources/techlib/report/novos/pdfs/stabiliza.pdf)** thus a high annealing temperature is favorable for good cold formability. The authors also reported that the r_m-value of the steels were higher that had lower solute carbon in the hot bands **(Storojeva et al., 2000)**. This is confirmed **(Figure 34)** by just comparing the Ti+Nb containing steels, where the r_m-values of the steels with the high solute carbon in hot strip (12-14 ppm) were remarkably lower than those of the steels without any solute carbon in hot strip. However, the r_m-value of Ti-free, just Nb-containing steel with the high solute carbon content (14 ppm) in the hot strip was almost as high as in the Ti+Nb steels without any solute carbon. Thus, titanium free IF steel exhibits a lower recrystallization start temperature and by this means enhances r_m-value in the final product, allowing compensation of the negative effect of solute carbon in the hot band. **Figure 35 (http://www.cbmm.com.br/ portug/sources/techlib/report/novos/pdfs/stabiliza.pdf)** summarizes test results of the r_m-value and BH-effect for annealing temperatures up to 840°C. It indicates, that a BH-effect >30 MPa together with r >1.7 can be obtained with the Ti-free, just Nb containing steel.

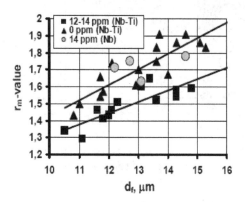

Fig. 34. r_m- value of steels with various solute carbon in hot band of IF steels.

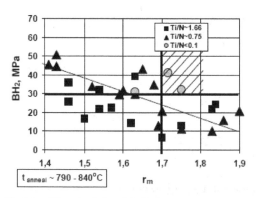

Fig. 35. r_m-value – BH-effect for Ti containing IF steels.

9. Recrystallization texture and drawability of warm rolled steels

The demand for thin and ultra thin rolled products led to an intensive search for alternatives to conventional hot and cold rolling processes. A number of studies were undertaken to investigate the advantages of rolling at temperatures between 850 and 500°C that is known as warm rolling (**Harlet et al., 1993**). Because of the lower reheat temperatures, warm rolling leads to lower production costs than hot rolling and requires significantly lower loads than cold rolling (**Harlet et al., 1993**). The latter factor means that higher reductions per pass can be produced by warm rolling than by cold rolling due to the lower plasticity of steel at room temperature. In conventional hot rolling, the finish rolling temperature is above the Ar3 temperature and in warm rolling finish rolling is made in the ferrite phase.

In the past years many researchers have worked on low, ultra carbon steels (**Sakata et al., 1997**) using warm rolling. The products obtained by ferritic hot rolling can be divided into two kinds according to the coiling temperature (**Mao, 2004**). One is a thin gauge soft and ductile hot rolled strip obtained by high temperature coiling for direct application that could be considered as a substitute for the conventional cold rolled and annealed sheet, and the other is a strained thin gauge hot strip gained by low temperature coiling for cold rolling and annealing, during which recrystallization texture strengthens by accumulating strains from hot rolling and cold rolling reductions.

Sa´nchez-Araiza et al (**Sa´nchez-Araiza, 2006**) reported on the texture changes in low carbon steel during recrystallization and established the nucleation and growth mechanisms applicable to warm-rolled quantitatively.

The requirement necessary for the development of (111) texture is the accumulation of strain in the matrix and to achieve uniform accumulation of strain through the thickness of sheet steel, lubricant is required in addition to the optimization of Ti and Nb concentration (**Figure 36**) (**Sakata et al., 1997**). In 1996, Kawasaki Steel Corporation constructed a new hot strip mill in the Chiba works, where sheet bars are welded between the coil box and the finish mill to accomplish fully continuous rolling. This 'endless hot strip mill' makes lubricant rolling practical. The r_m-value achieved using warm rolling technique is 2.9. This value is noteworthy since the best value obtained in the conventional process does not exceed 2.6.

Wang et al (**Wang et al., 2007**) studied drawability of Ti-stabilized IF steel by finish rolling the steel at 760ºC in the ferritic region with lubrication and coiling at 740 and 400ºC. The evolution of texture for both the high and low temperature coiling is shown in in **Figures 37-40**. The optical micrographs of the test steels in hot rolled and high temperature coiled status, cold rolled as well as annealed one are shown in **Figure 37**. It can be seen from **Figure 37(a)** that after high temperature coiling, the deformed microstructure vanished and the recrystallization microstructure is characterized by uniform and equiaxed grains. **Figure 37(b)** shows that after cold rolling, the grains can not be discerned and the obvious characteristics of the cold rolled microstructure is the formation of the in-grain bands denoted by the arrow. As shown in **Figure 37(c)**, deformed microstructure disappears and there are small and elongated grains after annealing. In order to obtain more equiaxed grains, the annealing temperature or the annealing time should be increased.

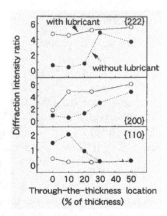

Fig. 36. Effect of lubricant on texture distribution through the thickness in ferrite rolled sheet steel; with finish rolling temperature 700 °C, cold reduction 50%, and annealing time 20s at 850° C (**Sakata 16 et al., 1997**)

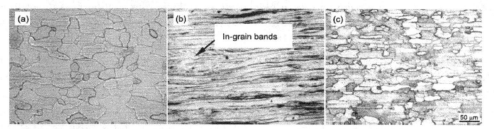

Fig. 37. Optical microstructures of samples in the condition of high temperature coiling: (a) hot rolled (b) cold rolled and (c) annealed samples (**Wang et al., 2007**).

Figure 38 shows the ϕ=45° ODF sections for high temperature coiled hot band, cold rolled and annealed sample. It is clear that after ferritic rolling and high temperature coiling, the most prominent texture intensity is along the γ-fiber and the maximum is at {111}<112>. Moreover, its characteristics are the same as that of the annealed texture in the condition of low temperature coiling, indicating that the hot band after high temperature coiling can be considered as a substitute for the conventional cold rolled and annealed sheet (**Wang et al., 2007**).

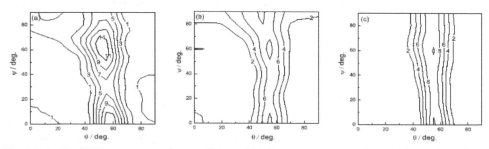

Fig. 38. ϕ=45º ODF sections in the condition of high temperature coiling: (a) hot rolled, (b) cold rolled and (c) annealed samples (**Wang et al., 2007**).

The optical micrographs of the low temperature coiled hot band, cold rolled as well as annealed one are shown in **Figure 39**. It is evident that a completely deformed microstructure is produced after hot rolling and low temperature coiling. Straighter grain boundaries and thinner deformation bands form after cold rolling. After annealing, the ferrite grains recrystallize completely, and small and uniform grains develop. **Figure 40** shows $\phi2=45°$ODFs of low temperature coiled hot band, cold rolled and annealed samples. The texture of hot band includes a strong α-fiber whose peak is at {001}<110> as well as a weak γ-fiber whose main component is {111} <110>. The components in the α-fiber intensify and the intensity of {111}<112> in the α fiber changes little after cold rolling. A complete γ-fiber with the peak at {111}<112> develops and the components in γ-fiber weaken evidently after annealing.

Fig. 39. Optical microstructures of samples in the condition of low temperature coiling: (a) hot rolled (b) cold rolled and (c) annealed samples (**Wang et al., 2007**).

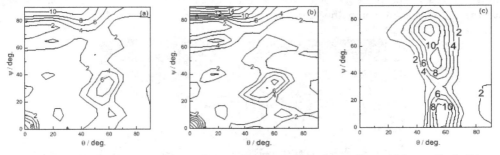

Fig. 40. $\phi=45°$ ODF sections in the condition of low temperature coiling: (a) hot rolled, (b) cold rolled and (c) annealed samples (**Wang et al., 2007**).

In a study on the effect of chemical composition and ferritic hot rolling on the formation of texture in the hot rolled and coiled, cold rolled, and cold rolled and annealed Ti and Ti+Nb added IF steels, it was reported that high r_m-values could be obtained at the same intensity levels of {111}<110> texture component if the grain growth was inhibited by a suitable addition of microalloying elements. (**Tiitto et al., 2004**). While as per the conventional wisdom it is known that grain growth during annealing is beneficial for drawability as it increases the volume fraction of grains with the {111} texture, leading to higher r_m-value. It was claimed that due to the grain refining effect of Nb, a small addition of the element (100 ppm) to the Ti alloyed IF steel increased the r_m-values in the annealed condition by about 15% at the same {111}<110> intensity levels. Thus, ferritic hot rolling seemed to be beneficial only if it contributed to the development of a strong intensity of the γ -fiber texture in association with a uniform and small grain size during annealing. Further they added, hot deformation of Nb and Ti alloyed steels at a high temperature (870°C) in the ferritic region led to higher {111}<110> intensities and higher r- values in the annealed condition than hot deformation at a lower temperature (800°C).

The development of the γ-fibre in annealed steels was linked (**Barnett & Jonas, 1997a; Duggan et al., 2000**) to the presence of a high volume fraction of grains containing shear bands after warm rolling. These bands appear to be the nucleation sites for recrystallised grains of the desirable orientation. After warm rolling, many of the grains in interstitial-free (IF) steels contain such shear bands and the steels then exhibit good forming characteristics after annealing (**Barnett, 1998**). By contrast, in low carbon steels, the presence of carbon in solid solution leads to dynamic strain aging during warm rolling and to high positive rate sensitivities. The latter prevent the formation of high densities of in-grain shear bands, leading to a lack of nucleation sites for the γ-fibre during annealing.

Timokhina et al (**Timokhina et al., 2004**) studied the effect of in grain shear bands on the volume fraction of favourable γ-fibre in IF, low carbon (LC) and LC with Cr, P and B added steels. Shear bands are usually contained within single deformed grains and are tilted by 20–35° with respect to the rolling plane. There are four types of in-grain shear bands: long (5-40 μm long--±15-40° to rolling direction), short (0.5-15 μm long), intense short (0.4-7 μm, ±5-45°) and intense long (continuous wavy lines--±15-40° to rolling direction) (**Barnett & Jonas, 1997b**) as shown in **Figure 41**. All the grains containing shear bands were characterized by zones of grain boundary displacement or stepping that provide evidence for local flow along the bands (**Barnett, 1996**). The presence of moderate amounts of long shear bands in IF steels was attributed to the formation of the γ-fibre after annealing (**Barnett & Jonas, 1997a**).

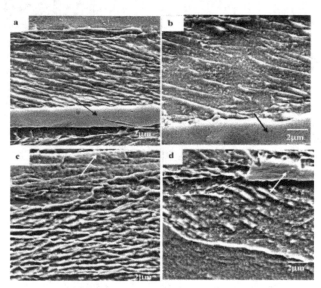

Fig. 41. SEM micrographs of (a) long shear bands (b) short shear bands, (c) intense long shear bands and (d) intense short shear bands. Arrows indicate the displacement zones (**Barnett & Jonas, 1997b**).

In their study, Timokhina et al (**Timokhina et al., 2004**) reheated the sample to 1050°C for 30 mins and subsequently warm rolled by 65% in a single pass pilot mill. Temperatures of 640°C and 710°C and average strain rates of 30s⁻¹ were employed followed by quenching. To establish grain orientation with different sheer bands, the grains were marked with hardness tester prior to electron back scattered diffraction analysis. The addition of alloying elements was observed to affect different types of shear band and the formation of γ -fibre. The IF steel was characterized by the largest number of grains with long shear bands with γ -fibre, 65 % and 67 % in the steels rolled at 640°C and 710°C respectively. Increasing the amount of carbon and decreasing the grain size, as in the LC steel, increased the proportion of grains with the short shear bands. Further, the proportion of grains containing intense long shear bands was higher in the LC steel than in the IF grade. The addition of carbon perceptibly decreased the volume fraction of grains with γ-fibre. The addition of chromium led to the formation of similar volume fractions of grains with short, long and intense long shear bands at both rolling temperatures. This led to an increase in the number of grains with beneficial γ -fibre to 46–47% compared to 25–26 % in the LC steel. However, in a subsequent study Pereloma et al (**Pereloma et al., 2004**) reported that although the formation of chromium carbides in the microstructures of the LC-Cr and LC-Cr+P steels removed carbon from solid solution and in this way slightly increased the fraction of γ-fiber nuclei formed at shear bands, the effect led to a reinforcement of the ND component (γ-fiber) only in the early stages of recrystallisation. The strength of γ-fiber gradually deteriorated with the progress of recrystallisation. The reasons put forward for this undesirable development were i) the retarding effect of the carbides on the mobility of grain boundaries during growth, ii) the influence of the particles on nucleus rotation during annealing and iii) the absorption of γ-fiber nuclei by other components during grain coalescence and growth. Thus, further study of the texture behaviour during grain coalescence and growth was required.

The addition of boron suppressed the formation of long, short and intense long shear bands and assisted in the formation of intense short shear bands that, in turn, decreased the volume fraction of grains with γ -fibre. The addition of phosphorus, on the other hand, increased the long and short shear band frequency and the proportion of grains with γ - fibre. In the Cr–B–Ti modified steel, the short, long and intense long shear bands were absent and they were replaced by intense short shear bands.

Jing et al (**Jing et al., 2011**) analyzed the grain boundary and microtexture characters of a rephosphorised high strength IF-steels under 700 and 800°C warm-rolled temperatures to observe the effect on deep drawability. It was found that while the samples were rolled at 700°C, more γ-fiber texture components, {111}<112>, {111}<110>, {554}<225>, low angle and CSL grain boundaries were formed that were beneficial for deep-drawability. However, the samples that were rolled at 800°C, manifested more α-fiber texture components and high angle grain boundaries that led to inferior deep drawing property. The average r-value was 1.32 for the samples rolled at 700°C and 1.05 for those that were rolled at 800°C, respectively.

Recrystallization texture investigation for IF-Ti and Ti stabilized IF-HS was conducted by Wang et al (**Wang et al., 2006**) under ferritic hot rolling and high-temperature coiling. Comparing with the completely recrystallized textures of the ordinary IF steel, the textures of the high-strength IF steel were of deformation type. This was attributed to the high phosphorous content in the high-strength IF steel that prevented recrystallization during the coiling process. For the ordinary IF steel, the texture components were mainly very weak rotated cube component {001}<110>at the surface, and partial α-fiber with key orientation {223}<110>orientation and <111>||ND texture at the mid-section and 1/4-section. While, for the high-strength IF steel, the texture components were orientation (Goss) at the surface and a sharp α-fiber extending from {001}<110> to {223}<110⟩ in association with a Weak <111>||ND texture at the mid-section and 1/4-section.

In another work by Ferry et al. (**Ferry et al., 2001**) on ultra low carbon steel (0.0036C, 0.03Ni, 0.019Mn, 0.03Al, 0.004Ti and 0.003N) it was found that the hot deformation microstructure had a strong influence both on the kinetics of recrystallization and texture development during cold rolling and annealing. In particular, a warm-deformed ferrite microstructure (lower finish deformation temperature (FDT)) recrystallized most rapidly to produce a strong <111>||ND recrystallization texture (γ -fibre) as it produces as-strained α in combination with the additional strain by cold rolling, which result in rapid recrystallization. while an initial coarse-grained ferrite microstructure recrystallized most sluggishly to produce a strong {001}<110> texture due to copious nucleation of grains at shear bands, which is consistent with previous studies (**Hutchinson, 1984; Muljono, 2001**). **Figure 42** (**Ferry et al., 2001**) shows φ_2=45° ODF sections in the fully recrystallized cold rolled and annealed samples following hot deformation at three significant finish rolling temperatures: (a) 920°C (fine, equiaxed ferrite), (b) 850°C (coarse ferrite) and (c) 600°C (warm deformed ferrite). The maximum intensity of the two most dominant recrystallization texture components, {001}<110> and {111}//ND, as a function of FDT are given in **Figure 43** (**Ferry et al., 2001**). It can be seen that the development of the strongest {111}//ND CRA texture is favoured when FDT is: (i) greater than 870°C (which produces fine-grained ferrite by transformation), and (ii) below, 800°C (which also produces fine-grained ferrite but with an additional true strain of 0.8 prior to cold rolling). Thus, it is indicated that warm rolling has a significant influence on final texture after cold rolling and annealing and warm rolling is capable of strengthening the {111}//ND (γ-fiber) recrystallization texture that is the favourable texture for drawability in the production of formable ultra low carbon steel sheets.

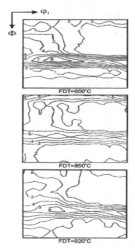

Fig. 42. φ_2=45° sections in Euler space showing recrystallization textures of the ultra low carbon steel with FDTs 600°C, 850°C and 920°C (contours: 1, 2, 3...random) (**Ferry et al., 2001**).

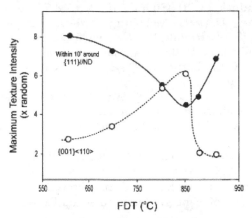

FDT (°C)

Fig. 43. Effect of FDT on maximum texture intensity of {001}<110> and within 10° around
<111>||ND (γ -fibre) of the annealed ultra low carbon steel (the maximum intensity is taken
from the each calculated ODF) (Ferry et al., 2001).

10. References

Auburn, Ph. & Rocquet P. (1973). Mem. Sci. Rev. Met., Vol. LXX, No. 4, p. 261.

Baker L. J., Daniel S. R & ParkerJ. D. (2002). Mater. Sci. Technol. Vol. 18, No. April, pp. 355-
367.

Banerjee K., Verma A. K. & Venugopalan T.(2008). Metall. and Mater. Trans. A, Vol. 39, No.
June, pp. 1410-1425.

Barnett M. R. & Jonas J. J. (1997). ISIJ Int., Vol. 37 pp. 697-705.

Barnett M. R. & Jonas J. J. (1997). ISIJ Int., Vol. 37, pp. 706-714.

Barnett M. R. & Jonas J. J. (1999).ISIJ Int., Vol. 39 pp. 856-873.

Barnett M. R. (1996).Ph.D. Thesis, McGill University, p. 52.

Barnett M. R. (1998). ISIJ Int., Vol. 38 pp. 78-85.

Barnett M.R. & Kestens L.(1999). ISIJ Int., Vol. 39, pp. 923-929.

Barrett M.R. & Jonas J.J.(1997) , ISIJ Int., Vol. 37, pp. 706-714.

Beranger G., Henry G & Sanz G. 1996.The book of Steel, Springer-Verlag, USA, 2-85206-981-
18 (1994), USA, pp. 935-951.

Cao W. D. , Lu X. P. , Sprecher A. F. & Conrad H. (1990). Mater. Lett., Vol. 9, p. 193.

Cao S. Q. (2005). Grain Boundaries and the Evolution of Texture in Interstitial-free (IF)
Steels, Ph.D. Thesis, Shanghai Jiao Tong University.

Chung J. , Era H. & Shimizu M. (1987). Metall. Trans. A, Vol. 18, No. June , pp. 957-968.

Conrad H. , Guo Z. , Sprecher A. F. (1989). Scripta Metall .Vol. 23, pp. 821-823.

Doherty R.D. (1985). Scr. Metall., Vol. 19, pp. 927-930.

Doherty R.D. , Hughes D.A. , Humphreys F.J. , Jonas J.J., Juul Jensen D., Kassner M.E.,
King W.E., McNelley T.R. McQueen H.J. & Rollett A. D. (1997). Mater. Sci. Eng. A,
Vol. 238, pp. 219–274.

Duggan B. J. , Liu G. I. , Ning H. , Tse Y. Y. (2000). Proc. Int. Conf. Thermomechanical
Processing of Steels, IOM Communications Ltd. London, UK, pp. 365 - 371.

El-Kashif E., Asakura K. & Shibata K. (2003). ISIJ Int., Vol. 43, No. 12, pp. 2007-2014.

Ferry M., Yu D. & Chandra T. (2001). ISIJ Int., Vol. 41, No. 8, pp. 876–882.

Gupta I. and Bhattacharya D. (1990). Metallurgy Of Vacuum Degassed Steel Products, ed. R. Pradhan, TMS, Warrendale, PA, TMS, pp 43– 72.

Harlet P. , Beco F. , Cantinieaux P. , Bouquegneau D. , Messien P. & Herman J. C. (1993). Int. Symp. on Low C Steels for the 90's, eds. R. Asfahani & G. Tither, TMS-AIME, Warrendale, PA, p. 389.

Hatherly M. & Hutchinson W. B. (1979). An introduction to textures in metals, The Institution of Metallurgists, Monograph 5.

He C.S. , Zhang Y.D., Wang Y.N. , Zhao X. , Zuo L. & Esling C. (2003). Scripta Mater., Vol. 48, pp. 737–74.

Hebert V., Louis P. , Zimmer P. & Delaneau P. (1992), CESSID document 92154.

Heckler A. J. & Granzow W. G. (1970). Metall. Trans A, Vol. 1, pp. 2089-94.

Held J. F. (1965). Mechanical Working and Steel Processing IV, ed.D. A. Edgecombe, American Institute of Mining, Metallurgical and Petroleum Engineers, New York, p. 3.

Holie S. (2000). Mater. Sci .Technol., Vol 16, pp. 1079-1093.

Hölscher M. , Raabe D. & Lu"cke K. (1991). Steel Res., Vol. 62, No. 12, pp. 567-75.

Hook R. E. (1990). Metallurgy of vacuum-degassed steel products (ed. R. Pradhan), 1990, Warrendale, PA, Metallurgical Society AIME, p.263.

Hosford W. F. & Backholen W.A. (1964). Fundamentals of deformation processing, Syracuse, Press, New York. P. 259.

Hu H. & Goodman S.R. (1970). Metall. Trans., Vol. 1, pp. 3057-64.

Hu H. (1977). Metall. Trans. A, Vol. 8, pp. 1567-75.

Hughes I.F. & Page E.W. (1971). Metall. Trans., Vol. 2, pp. 2067-75.

Hutchinson W. B. (1984). Intl. Mater. Reviews, Vol. 29, No. 1, pp. 25-42.

Hutchinson W.B. & Ushioda K. (1984). Scand. J. Met., Vol. 13, p. 269-284.

Irie T. , Hashiguchi H. , Satoh S. , Konoshi M. , Takahashi K. & Hashimoto M (1981).Trans. Iron Steel Inst. Jpn., Vol. 21, No. 11, pp. 793-801.

Irie T. , Satoh S. , Yasuda A. & Hashimoto O. (1982). Metallurgy of Continuously Annealed Sheet Steel, TMS, Warrendale, PA, TMS, pp. 155-171.

Jeong W. C. (2000). Metall. and Mater. Trans. A, Vol. 31, No. April, pp. 1305-1307.

Jing C. , Wang M. , Liu X. , Tan Q. , Wang Z. & Han F. (2011). Mater. Sci. Forum, Vol. 682, pp. 71-74.

Juntunen P. , Raabe D., Karjalainen P. , Kopio T. & Bolle G. (2001). Metall. & Mater. Trans. A, Vol. 32, No. Aug, pp. 1989-95.

Kang H. , Garcia C. I. , Chin K. & DeArdo A. J. (2007). ISIJ Int., Vol. 47 No. 3, pp. 486–492.

Katoh H. , Takechi H. , Takahashi N. & Abe M. (1985). Int. Conf. on Technology of Continuously Annealed Cold Rolled Sheet Steels, ed. Pradhan. R, Proc. . TMS-AIME, Warrendale, PA, USA, pp. 37-60.

Kawasaki K., Senuma T. & Sanagi S. (1991). Processing, Microstructure and Properties of Microalloyed and other Modern HSLA Steels, ISS, Warrendale, PA, pp. 137 - 144.

Kestens L.., Jonas J. J., Van Houtte P. & Aernoudt E. (1996). Textures and Microstructures, Vol. 26-27, pp. 321-335.

Kim S., Choi I., Park I. & K. Cho (2005). Mater. Sci. Forum, Vol. 475-479, pp. 475-479.

Kitamura M. , Tsukatani I. & Inoue T.(1994) : ISIJ Int., Vol. 34, No. 1, pp. 115 – 122.

Klein A. J. & Hitchler, E. W. (1973). Met. Eng. Q., Vol. 13, pp. 25 – 27.

Kojima N. , Mizui N.& Tanioku T. (1993). Sumitomo Search, Vol. 45, No. 5, pp.12 – 19.

Kozeschnik E. , Pletenev V. , Zolotorevsky N. & Buchmayr B. (1999) , Metall. & Mater Trans. A, Vol. 30, No. June, pp. 1663-1673.

Krauss G., Wilshynsky D. O. & Matlock D. K. (1991). Interstitial Free Steel Sheet: Processing, Fabrication and Properties, eds. L. E. Collins and D. L. Baragar, CIM/ICM1, Ottawa, pp. 1- 14.

Lankford W. T. , Snyder S. C. & Bauscher J. A. (1950). Trans.AS1/I, Vol. 42, pp.1197 – 1232.

Lebrun J. L., Maeder G. & Parniere P.(1981) : Proc. 6th Intl. Conf on Texture of Materials, Vol. 2, Tokyo, The Iron and Steel Institute of Japan, p. 787.

Mao X. (2004). Iron Steel, Vol. 39, No. 5, p. 71.

Martin, J.W., Doherty R. D. , Cantor B. (1997). Stability of Microstructure in Metallic Systems (2nd edition)., Cambridge University Press, Cambridge.

Matlock D. K., Allan B. J. & Speer J. G. (1998). Proc. Conf. Modern LC and ULC Sheet Steels for Cold Forming Processing and Properties, ed. W. Bleck, Aachen, Verlag Mainz, pp. 265 - 276.

Matsudo K. , Osawa K. & Kutihara K. (1984). Technology of Continuously Annealed Cold-Rolled Sheet Steel, ed. R. Pradhan, TMS-AIME, Michigan, pp. 3-36.

McQueen H.J. , Rollett A.D. (1997). Materials Science and Engineering, Vol.A238, pp. 219–274.

Mendoza R. , Huante J. , Alanis M., Gonzalez-Rivera C. & Juarez-Islas J. A. (2000). Mater. Sci. Eng A, Vol. 276, pp. 203-209.

Mendoza R. ,Alanis M. , Aramburo G. , Serrania F. & Juárez-Islas J.A. (2004). Mater. Sci. Eng.A, Vol. 368, pp. 249–254.

Meyzaud Y.& Parniere P.(1974). Mem. Sci. Rev. Met., Vol. LXXI, No. 7-8, pp. 423.

Mizui N. , Okamoto A. & Tanioku T.(1990). Proc. LTV/SMI Technology Exchange Meeting, Ltv Steel/Sumitomo Metal Industries.

Muljono D. , Ferry M. & Dunne D.P.(2001). Mater. Sci. Eng. A, Vol. 303, pp. 90–99.

Nishimoto A. , Inagaki J. & Nakaoka K.(1982).Tetsu-to-Hagne, Vol. 68, pp.1404-1410.

Ono S. , Shimomura T., Osawa K. & Matsudo K.(1982). Transaction ISIJ, Vol. 22, pp. 732-738.

Park Y.B. , Kestens L.& Jonas J.J. (2000). ISIJ Int., Vol. 40, pp. 393-401.

Pereloma E. V. , Timokhina I. B. , Nosenkov A. I. & Jonas J. J. (2004). Metallurgija, Vol. 43, No. 3, pp. 149-154.

Perera M., Saimoto S. & Boyd D. (1991). Interstitial Free Steel Sheet: Processing, Fabrication and Properties, eds.. L. E. Collins and D. L. Baragar, ;Ottawa, CIM/ICM, pp. 55 - 64.

Ray R. K. & Jonas J. J. (1990). Int., Mater. Rev., Vol. 35, No. 1, pp. 1-36.

Ray R. K., Jonas J. J. & R. E. Hook (1994). Intl. Mater. Reviews, Vol. 39, No. 4, pp. 129-172.

Rege J. S. , . Garcia C. I & DeArdo A. J.(1997). Proc. 39th Mechanical Working and Steel Processing, Vol. 35, ISS, Warrendale, PA, USA, pp. 149-158.

Rege J. S., Hua C., Garcia I. & DeArdo A. J. (2000). ISIJ Intl., Vol. 40, No. 2, pp.191-199.

Ruiz-Aparicio L.J. , Garcia C.I. & Deardo A.J. (2001) , Metall.and Mater. Trans. A, Vol. 32, No. September, pp. 2325-2334.

Sa´nchez-Araiza M. , Godet S. , Jacques P.J. & Jonas J.J. (2006). Acta Mater. , Vol. 54, pp. 3085–3093.

Sakata K. , Matsuoka S. , Obara T. , Tsunoyama K. & Shiraishi M.(1997). Materia. Japan, Vol. 36 No. 4 p. 376.

Samajdar I. (1994).Ph.D. Thesis, Drexel University.

Sarkar B., Jha B. K. & Deva A. (2004). J. Mater. Eng. and Perform., Vol. 13, No. 3, pp. 361-36

Schulz L. G. (1949). J. Appl. Phys. Vol. 20, No.11, pp.1030-33.

Song X., Yuan Z. Jia J. , Wang D. , Li P. & Deng Z (2010). J. Mater. Sci .Technol. Vol. 26, No. 9, pp.793-797.

Storojeva L. , Escher C. , Bode R. , K. Hulka & Yakubovsky O. (2000). IF Steels 2000, ISS, Warrendale, PA, p. 289.

http://www.cbmm.com.br/portug/sources/techlib/report/novos/pdfs/stabiliza.pdf

Takahashi M., Okamoto, A. (1974). Sumimoto Met., Vol. 27,pp. 40-49.

Tanioku T. , Hobah Y. , Okamoto A. & N. Mizui (1991). SAE Technical Paper 910293, Society of Automotive Engineers, Warrendale, PA, USA.

Tiitto K. M. , Jung C. , Wray P. , Garcia C. I. & DeArdo A. J. (2004). ISIJ Int., Vol. 44, No. 2, pp. 404–413.

Timokhina I. B. , Nosenkov A. I. , Humphreys A. O. , J. J. Jonas & Pereloma E. V.(2004). ISIJ Int., Vol. 44, No. 4, pp. 717–724.

Tither G. & Stuart H. (1995). HSLA Steels '95', ed. L. Guoxun et al.., Chinese Society for Metals, Beijing, pp. 22-31.

Tither G., Garcia C. I., Hua M. & Deardo A. J. (1994). Int. Forum for. Physical Metallurgy in IF Steels, Iron and Steel Institute of Japan, Tokyo, pp. 293-322.

Tokunaga Y. & Yamada M. (1985). Method for the Production of Cold Rolled Steel Sheet Having Super Deep Drawability, US Patent 4,504,326.

Tokunaga Y. & Kato H. (1990). Metallurgy of Vacuum Degassed Products , TMS, Warrendale, PA, pp. 91 -108.

Tsunoyama K. (1998). Phys. Stat. Sol. (A), Vol. 167, No. 427, pp. 427-433.

Tsunoyama K. , Sakata K., Obara T. , Satoh S. , Hashiguchi K. & Irie T.(1988). Hot and Cold Rolled Sheet Steels, eds. R. Pradhan and G. Ludkovsky, TMS, Warrendale, PA, pp. 155 - 165.

Tsunoyama K., Satoh S., Yamasaki Y. & Abe H. (1990). Metallurgy of Vacuum Degassed Products, 1990, TMS, Warrendale, PA, pp. 127 -141.

Wang Y. N. , He C. S. , Zhao X. , Zuo L., Zhi Q. Z. & Liang Z. D (2000). Acta Metall Sinica, Vol. 36, No. 2, p. A126.

Wang Z. , Guo Y. , Xue W., Liu X. & Wang G.(2007). J. Mater. Sci. Technol., Vol. 23 No.3, pp. 337-341.

Wang Z.D. , Guo Y.H. , Sun D.Q. , Liu X. H. & Wang G.D. (2006). Mater. Charact., Vol. 57, No. 4-5, pp. 402-407.

Whiteley R. L. & Wise D. E (1962). Flat rolled products III, Interscience, New York, pp. 47 – 63.

Wilshynsky-Dresler D. O., Matlock D. K. & Krauss G. (1995): ISS Mech. Work. Steel Process. Conf., 1995, 33, pp. 927 – 940.

Wilson D. V. (1966). J. Inst. Met., Vol. 94, pp. 84 – 93.

Xiaojun G. & Xianjin W. (1995). Textures and Microstructures, Vol. 23, pp. 21-27.

Yasuhara E., Sakata K. , Furukimi O. & Mega T. (1996): Proc. 38th Mechanical Working and Steel Processing, Vol,. 34, Cleveland, Ohio, USA, pp. 409-415.

Yoshicla K. at. (1974): Deep Drawing Research Group, Proc. 8th Biennial IDDIG Congr. Gothenburg, 1974, pp. 258 – 268.

6

The Failure Mechanism of Recrystallization – Assisted Cracking of Solder Interconnections

Toni T. Mattila and Jorma K. Kivilahti
Aalto University
Finland

1. Introduction

The typical user environment load spectrum varies significantly between different electronic applications but changes in temperature are involved in nearly all of them. Owing to the increasing number of integrated high-performance functions, smartphones and handheld computers, for example, can experience significant changes in temperature during normal operation. The changes in temperature are typically caused either by internally generated heat dissipation or changes in the temperature of the environment. Today the maximum temperatures inside modern portable electronic products are in the range of 60-70 $^{\circ}$C but occasionally they can even rise above 90 $^{\circ}$C [1,2].

The thermomechanical reliability of electronic component boards has been one of the most studied aspects in the field for several decades. Sustained interest in this topic has endured primarily because: (a) the power densities and heat dissipation of novel electronic devices have increased; (b) electronic devices are being designed for use in ever harsher environments, such as the engine compartments of automobiles, and (c) newly developed materials whose long-term behavior in electronic applications is still unknown are continuously being introduced into electronic assemblies.

Thermomechanical strains and stresses in electronic assemblies are produced when the thermal expansion and contraction of materials is restricted. The standard thermal cycling tests extend the temperature range of electronic devices under normal operating conditions in order to accelerate the accumulation of failures. The maximum extreme temperatures in some of the standards are set to -65 $^{\circ}$C and +150 $^{\circ}$C, but the conditions of -45 $^{\circ}$C and +125 $^{\circ}$C with 15- to 30-minute dwell times are most commonly used [3-5]. The coefficients of thermal expansion (CTE) of most printed wiring boards (PWB) are much higher than those of most packages. For instance, the CTE of the most commonly used PWB base material, FR–4, is about 16-17 x 10^{-6}/$^{\circ}$C [6], whereas that of silicon is only 2.5 x 10^{-6}/$^{\circ}$C [7]. Furthermore, because of the large volume fraction of silicon in electronic packages (see Fig. 1a), packages have much higher rigidity than PWBs and, consequently, as the component boards are exposed to changes in temperature, strains and stresses are concentrated in the solder interconnections between the packages and the PWB, as the natural expansion/contraction of the PWB is restricted by the packages (see Fig. 1b). Therefore, the reliability of most electronic products under changing thermal conditions is determined by the ability of the solder interconnections to withstand thermomechanical loads. Thermomechanical strains

(a) BGA Package

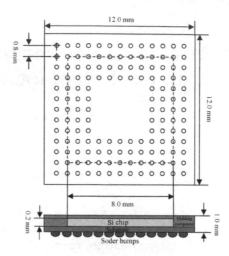

(b) Component Board Assembly

(c) Displacements under Change of Temperature

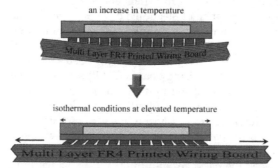

Fig. 1. a) A Ball Grid Array (BGA) package commonly used in high-density electronic devices; b) structure of the component board assembly; c) formation of strains and stresses under an increase in temperature.

and stresses in solder interconnections are built up to different extents, of course, at all length scales, ranging from the submicron intermetallic particles submerged in the tin matrix of the tin-rich solder interconnections to the structural dimensions of a product. However, as the CTEs of most printed wiring boards is many times higher than that of most packages, the influence of strains and stresses at higher length scales (i.e. the board or the product level) are more influential.

The mechanical properties of solder interconnections are, of course, dependent on the microstructures formed during soldering but the fact that they are not stable and tend to change distinctly during the operation of products makes the evaluation of reliability a challenging task. Thus, detailed understanding of the evolution of microstructures under thermal cycling conditions and their influences on the failure of solder interconnections is essential because it can provide us with the means not only to improve reliability but also to develop more efficient and meaningful methods for the reliability evaluation of and lifetime prediction.

Despite it being a popular topic of academic research for decades, we are only beginning to understand the complexities related to the failure mechanisms of solder interconnections under cyclic thermomechanical loading. Justification for this perhaps surprising statement lies in the fact that it was only recently that the widely used tin-lead solder alloys were replaced with new lead-free materials and this change has made a comprehensive re-assessment of reliability necessary [8,9]. It is well known that the reliability of solder interconnections is strongly influenced by the microstructures formed during soldering and their evolution during use, but it is particularly interesting to observe that the microstructural changes in tin-rich lead-free solders are markedly different from those observed in tin-lead solders, where failure takes place as a result of the heterogeneous coarsening of tin and lead phases and eventual cracking of the bulk solder (see, e.g., [10,11]). The microstructural observations of failed tin-silver-copper solder interconnections have indicated that the microstructures of solder interconnections change distinctly before cracking but by a different mechanism, namely cracking that is assisted by recrystallization (see, e.g. [12,15]).

In this chapter we discuss the failure mechanisms of tin-rich lead-free solder interconnections of electronic component boards from the perspective of the evolution of microstructures. The focus is placed on the identification of the factors driving the microstructural evolution in lead-free interconnections that creates the preconditions for the energetically feasible intergranular propagation of cracks through the interconnections. The microstrucutural approach to the reliability of electronics is useful, particularly because many failure mechanisms are related to the inevitable evolution of microstructures that takes place during the normal operation of products. Owing to the extensive research carried out over the last decade we know the microstructures and mechanical properties of many lead-free solder compositions well, but the evolution of microstructures during the operation of products has still gained little attention. However, before going into details of the changes in microstructures one should have a generalized understanding of the microstructures to be discussed when they are in an as-solidified state. Therefore we will begin this chapter with a brief overview of the solidification and microstructures of tin-rich lead-free solder interconnections formed during solidification. After that the restoration of solder interconnections and the conditions under which the recrystallization is initiated are discussed. The onset and progress of recrystallization in solder interconnections will be discussed in detail. The failure of the solder interconnections is discussed in a separate section to emphasize the fact that the propagation of cracks takes place after the change in microstructures as a result of recrystallization and that recrystallization significantly enhances their propagation in the solder interconnections. Finally, a method to predict the changes in microstructure and, ultimately, to predict the lifetime of solder interconnections is presented and discussed.

2. As-solidified microstructures of tin-rich solder interconnections

The majority of lead-free solders are based on tin (Sn), with a few alloying elements. Silver (Ag) and copper (Cu) are the most common major alloying elements but alloys with minor additions of elements such as nickel, antimony, indium, germanium, manganese, bismuth, zinc, or rare earth elements are also commercially available. However, for the sake of simplicity let us consider the three-component SnAgCu alloy with near-eutectic composition, which is the most commonly used solder alloy in the electronics industry.

The eutectic composition of the tin-silver-copper alloy is about Sn3.4Ag0.8Cu [13,14]. Nearly all of the solder compositions used in the electronics industry, such as the composition of Sg3.0Ag0.5Cu that is most widely used today, have silver and copper concentrations below the eutectic concentration. For such compositions, the tin-rich phase (i.e., high-tin solid solution) is formed first at the beginning of solidification and its morphology strongly affects the solidified microstructure. Owing to the high tin content of the near-eutectic SnAgCu alloys (more than 95 wt-% Sn), the solidification and generated microstructures of the interconnections are most significantly influenced by the solidification of the tin-rich phase, even though in practice the tin-rich phase may or may not always be the first phase to form during solidification. This is because the dissolution of the contact metallizations changes the nominal composition of the solder and can thereby influence solidification, as will be pointed out shortly.

Figures 2a-b show an example of a cross-section of a near-eutectic SnAgCu solder interconnection in the as-solidified condition as imaged by employing the optical (cross-polarized) microscopy and the scanning electron microscopy. The boundaries between the contrasting areas in Figure 2b are high-angle boundaries between the matrices of solidification colonies[1] (the orientation difference between adjacent regions is quite large, larger than about 15°). A uniformly oriented cellular solidification structure of tin is enclosed within the colony boundaries. There is a low-angle orientation difference between the cells enclosed by the high-angle colony boundaries. The cellular structure of tin is clearly distinguishable as cells are surrounded by eutectic regions (see Fig. 2c-d). It is also noticeable that the solder interconnections, such as that shown in Figure 2, are commonly composed of a few solidification colonies of relatively uniformly oriented tin cells. [12,15] Similar observation have been made by other groups also [16,17-21].

---

[1] We have chosen to use the term "solidification colony" in order to emphasize the fact that under the reflow conditions employed, a cellular structure is generated in which the difference in crystal orientations between individual cells is small (few degrees or less). The use of this term also helps us to make a verbal distinction between the as-solidified microstructures and the recrystallized grains.

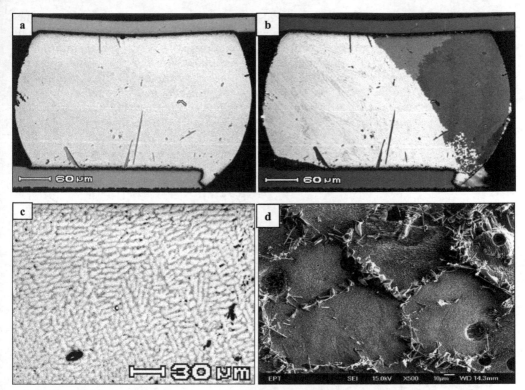

Fig. 2. The as-reflowed microstructures of a near-eutectic SnAgCu interconnection: (a) optical bright field image of a cross-section; (b) a cross-polarized light image of the same cross-section highlights the colony boundaries (high angle boundaries); (c) an SEM micrograph showing the cellular structure within a colony, where the cells are separated by low-angle boundaries; (d) the cell structure of a solidification colony is emphasized by the small intermetallic particles that surround the tin cells (the sample has been selectively etched).

Dendritic morphologies of the tin-rich phase are also reported in the literature; see e.g. [22-25]. It is evident that this phase can solidify in different morphologies depending on the solidification conditions (e.g., cooling rate and metallizations in contact with the solidifying solder) and the nominal composition of the solder. The volume of the solder interconnections obviously influences the microstructures formed during solidification, as a dendritic structure of the tin-rich phase is more often observed in the case of large solder volumes, such as cast dog-bone or lap-joint pull test specimens (prepared for the mechanical characterization of solders) or packages with a relatively large (≈ 1 mm) bump diameter (e.g. [16,26-28]). However, the as-solidified microstructure in the near-eutectic solder interconnections in our studies has consistently been cellular, regardless of the compositions of the near-eutectic SnAgCu paste or bump alloy, contact pad metallization, package type or dimensions, or the setup parameters of our full-scale forced convection reflow soldering oven.

In order to rationalize the formation of the observed microstructures, it is useful to examine the solidification of solder interconnections with the help of equilibrium phase diagrams. It should be kept in mind, however, that the equilibrium diagrams do not contain information about either the effect of the cooling rate or the morphology of the phases and, thus, the solidification structures are examined as equilibrium solidification[2]. Figure 3 presents the tin-rich corner of the SnAgCu phase diagram with the isothermal lines representing the liquidus temperatures, and the primary phase regions of the tin-rich phase, Cu_6Sn_5 and Ag_3Sn.

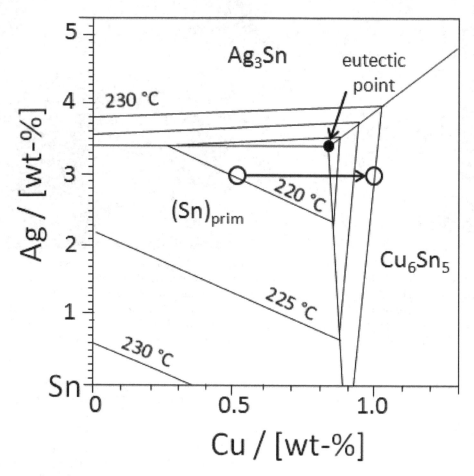

Fig. 3. Tin-rich corner of the SnAgCu phase diagram with isothermal lines and primary phase regions [29]. The arrow between the two circles represents the change in nominal composition (circle) owing to copper dissolution.

[2]For example, in order to take the undercooling (which will be discussed shortly) into consideration, one should extrapolate the liquidus surfaces of the Cu_6Sn_5 and Ag_3Sn phase to lower temperatures and lower (or remove) the liquidus surface of the $(Sn)_{prim.}$ phase.

Let us consider, for example, the composition of Sn3.0Ag0.5Cu, which is a very commonly used alloy in the reflow soldering of component boards. The solidification of this composition starts with the formation of the tin-rich solution phase when the interconnections are cooled down from the peak reflow temperature to below the liquidus temperature of about 220 °C. The secondary phase, namely Cu_6Sn_5 or Ag_3Sn, is formed only after the nominal composition of the remaining liquid meets the curve of two-fold saturation, after which the solidification of the interconnections proceeds by the binary eutectic reaction (liquid transforms to $(Sn)_{eut}$ + the secondary phase; see [15] for more details).

It should also be noticed that in practice the microstructures formed on other pad metallizations can differ notably, even though the same solder compositions are used (see, e.g., [15,30,31]). The dissolution of the contact pads or pad metallizations of packages and printed wiring boards during reflow changes the composition of the molten solder. The influence of contact metallizations depends primarily on the dissolution rate and reactivity of the metallizations. For example, the dissolution rate of copper in near-eutectic SnAgCu solder is about 0.07 µm/s, but that of Ni is more than an order of magnitude lower and can be considered negligible [15,32,33]. Thus, practically all nickel that is dissolved into liquid solder is used in the reaction to form intermetallic layers. However, in the cases where the solder is in direct contact with copper pads (i.e., no protective coating or organic soldering preservative is used on the copper soldering pads) the dissolution rate of copper from the pads into typical BGA solder interconnections (with a bump diameter of about 0.5 mm) is high enough to lift the copper concentration above 1 wt-% during soldering (see Fig. 3). This change in the nominal composition of the liquid can change the primary phase formed during solidification from the tin-rich phase to Cu_6Sn_5. Therefore, the as-solidified microstructures on the copper pads often show large amounts of primary Cu_6Sn_5 (hexagonal) tubes or rods in the microstructure that are absent from the microstructures of interconnections that are soldered on slow dissolution rate metallizations, such as nickel. Furthermore, interconnections soldered on copper pads, as opposed to those soldered on nickel, typically show more numerous and larger Cu_6Sn_5 particles embedded at the boundaries between the tin cells that are formed in the binary solidification, as the composition of the liquid moves a greater distance along the eutectic valley (along the curve of two-fold saturation).

There is an important consequence related to the increased copper content: the relatively large primary phase needles or particles dispersed in the solder interconnections can influence the evolution of solder interconnection microstructures. The non-coherent high-angle boundaries between the Cu_6Sn_5 crystals and tin matrix provide good nucleation sites for the recrystallization. It has been previously demonstrated that the second phase particles can accelerate the nucleation of recrystallization in common structural alloys [34,35]. Readers interested in particle-stimulated nucleation of recrystallization can refer e.g. to [36,37].

However, as pointed out earlier, in practice the solidification process departs somewhat from that of equilibrium solidification. Taking the undercooling into account would result in the nucleation of the Cu_6Sn_5 as a primary phase at even lower copper concentrations. Figure 4 illustrates an as-solidified microstructure of the Sn3.0Ag0.5Cu solder interconnection. It is interesting to observe that the primary Cu_6Sn_5 needles in the micrographs have the tendency to nucleate at the free surfaces of the molten interconnection, most probably on oxide

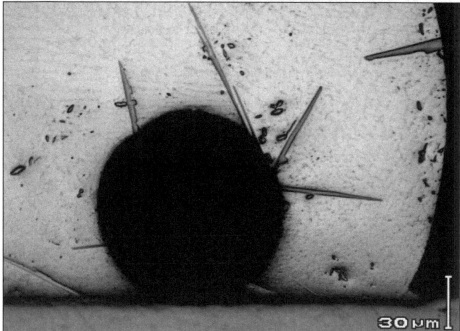

Fig. 4. As-solidified microstructure of a Sn3.0Ag0.5Cu solder interconnection illustrate how the primary Cu_6Sn_5 particles have nucleated on oxide particles on the surface of the liquid.

particles on the liquid surfaces, instead of the package or the PWB side interfaces. It should also be mentioned, without going into detail, that it seems as if the solidification of the tin-rich phase is controlled by the kinetics of heterogeneous nucleation at the surface of (Cu_6Sn_5) intermetallic layers [38]. Darveaux et al. observed experimentally that SnAgCu solder alloys with a higher copper concentration exhibit a higher amount of undercooling than those with a lower concentration [39].

Before we move on to the evolution of microstructures under operating conditions, we would like to point out a few aspects to consider in more detailed investigations of the failure mechanisms of recrystallization-assisted cracking of solder interconnections. It is particularly noteworthy that the Cu_6Sn_5 or the Ag_3Sn phase can nucleate with minimum undercooling in the liquid SnAgCu interconnections [40-42] but the nucleation of the tin-rich phase results in a significantly wide range of undercooling that can extend up to 60 °C [43-48]. The large amounts of undercooling indicate apparent difficulties in the nucleation of tin crystals in the liquid, which can be one of the reasons why there are very often only few orientations of the Sn-rich phase observed on a cross-section of solder interconnections. The tendency to form only a few large crystals, which can be several hundred micrometers in diameter, has been observed in interconnections of various length scales, ranging from about 100 µm (the diameter of a Flip Chip interconnection) to millimetre scale (lap-joint specimens used in material characterization) [15-20,28,49]. Furthermore, the fact that sometimes neighboring regions of a cross-section share a twinning relationship (indicated by a misorientation angle of about 60° between adjoining regions) suggests that these crystals originate from a common nucleus and, thus, the number of different crystals can be even smaller that the amount determined by the commonly employed qualitative method of cross-polarized light microscopy [16,20,21].

What has been stated above indicates clearly that mechanical behavior of solder interconnections is most probably quite different from that of a "normal" polycrystalline material. As has been pointed out and is currently being studied by many authors, the fact that the physical and mechanical properties of the tin-rich phase exhibit significant anisotropic behaviour[3] can cause severely non-homogeneous deformation and the formation of internal stresses in the solder interconnections [52-54]. As the internally produced strains and stresses are combined with the higher-level strains and stresses, as caused by the differences in the coefficients of thermal expansion of printed wiring boards and packages, it is clear that the thermomechanical response of the solder interconnections becomes very complex and unique to each solder interconnection. Furthermore, the grain boundary cracking should not occur in the as-solidified structure due to the absent of high angle boundaries (other than those between colonies). This can lead to unpredictable failure sites, as reported in [54,55]. Therefore, when stress is applied to interconnections having this kind of microstructure, they undergo microstructural evolution before fractures can propagate. Investigations of the microstructures of failed solder interconnections have indicated that the microstructures formed during solidification are not stable and will change notably during the operation of products [15,17,18,56-63].

---

[3] The coefficient of thermal expansion along the c-axis of the tetragonal unit cell {c/a ratio of about 0.5} is twice that along the other two axes {a and b-axis}; the elastic modulus along the c-axis is only about 0.6 times that along the other two axes [[50],[51]]

3. The role of recovery and recrystallization in the failure mechanisms of solder interconnections

The reliability of electronic devices is commonly assessed by employing standard thermal cycling tests that place the extreme temperatures in the range of about - 45 °C to +125 °C. The thermomechanical stresses formed in the solder interconnections under these conditions are high enough to cause instantaneous plastic deformation of the commonly used near-eutectic SnAgCu solders [63]. Furthermore, this whole temperature range remains above the 0.3-0.4 homologous[4] temperature range of the solders, which is the temperature range above which the time-dependent deformation of metals becomes significant. Thus, time spent at either elevated or lowered temperatures allows diffusion creep processes to transform the elastic strain part of the total strain into inelastic strain[5]. The energy stored during deformation acts as the driving force for the evolution of the microstructures.

The initiation of microstructural changes in solder interconnections is localized because of the highly non-uniform distribution of strains inside the solder interconnections. Figure 5 shows the calculated strain energy density distribution in the cornermost solder interconnection of the component board assembly shown in Figure 1. It should be pointed out that there is a difference in the rate of strain energy accumulation during thermal cycling between the solder regions on the opposite sides of the interconnections and, therefore, changes in the microstructure are observed first on the package side regions of the interconnections, where the inelastic deformation is more extensive.

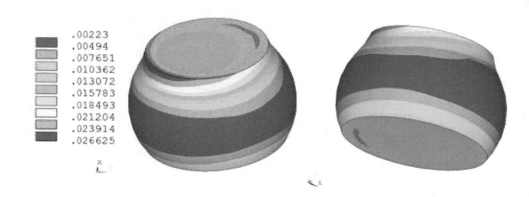

Fig. 5. Calculated strain energy density of the cornermost solder interconnections of the component board assembly sketched in Figure 1 [63,64].

---

[4]Defined as the ratio of the prevailing temperature to the melting point of a solvent metal; both expressed in Kelvin.

[5]Here we consider 'total strain' = elastic + inelastic strain = elastic + plastic + creep strain.

In studies reported in more detail elsewhere, the evolution of microstructures was investigated as a function of thermal cycles by taking out samples at fixed intervals during the course of the test and inspecting them for the development of microstructures and failures [64,65]. The results showed that the evolution of microstructures in the strain concentration regions commenced with gradual evolution of the cellular solidification structure, but after some time, i.e., the incubation period, the microstructures changed discontinuously by recrystallization. A similar observation has been reported in [66].

3.1 Restoration of tin-rich solder alloys

When solder interconnections are deformed plastically, a part of the work is stored in solders as lattice defects, mainly in the strain fields of dislocations. The increased internal energy of deformed solder acts as the driving force for the competing restoration processes, recovery and recrystallization. It is well known that the degree of restoration by recovery depends on the stacking fault energy of the solder alloy. At the time of writing there is little information in the literature about the recrystallization behavior of tin-based solder alloys but since the near-eutectic SnAgCu alloy contains more than 95 wt-% of tin, recrystallization studies on pure tin can be considered indicative, bearing in mind that the alloying elements in solid solutions, as well as small precipitates, do affect the restoration processes. Creep studies carried out with high-purity tin have suggested that the stacking fault energy of tin is high [67,68]. The recovery is very effective in high-stacking fault energy metals, such as aluminum and iron, as a result of the efficient annihilation of dislocations by cross slip and climb. Therefore one can expect the restoration of high-tin solder alloys to take place to a large extent by recovery. Gay et al. and Guy have observed that pure tin (99.995% purity) recrystallizes at room temperature even after a modest deformation (reductions of a few percent) [69,70]. However, in a more recent study Miettinen concluded that that even highly deformed (up to a 50% reduction) near-eutectic SnAgCu solders do not recrystallize statically when annealed at 100 °C after deformation at room temperature [71]. Korhonen et al. also failed to observe recrystallization in dynamic fatigue tests performed at room temperature [72]. All these results indicate that recovery is effective also in near-eutectic SnAgCu solder alloys. Because recovery and recrystallization are competing processes, the progress of recovery can reduce the driving force of recrystallization significantly and recrystallization may not always initiate. On the other hand, it is well documented that near-eutectic SnAgCu interconnections do recrystallize under dynamic loading caused by changes in temperature (between -45 °C and +125 °C), as well as under power cycling conditions (between room temperature and +125 °C) [15,17,18,56-63,73]. Thus, it seems that near-eutectic SnAgCu solder interconnections recrystallize only under restricted loading conditions: dynamic loading conditions where the strain hardening is more effective than the recovery.

Figure 6a shows a micrograph of the recrystallized microstructure on the package side neck region of a thermally cycled interconnection taken by employing optical microscopy and cross-polarized light. Figure 6b shows an electron backscatter diffraction orientation map of the same surface. The black lines in the figure represent the boundaries where the crystal orientation of the adjacent grains exceeds 30° and correspond well to the grain boundaries visible in the optical micrograph in Figure 6a.

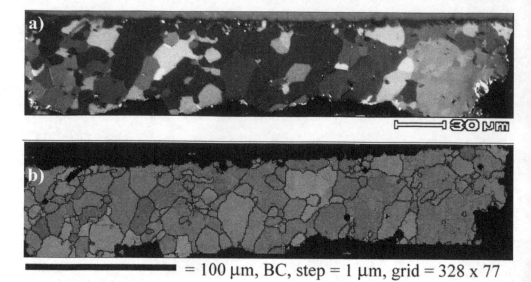

= 100 μm, BC, step = 1 μm, grid = 328 x 77

Fig. 6. a) Optical micrograph showing the recrystallized structure on the package side interfacial region of the SnAgCu solder interconnection taken with polarized light; b) EBSD graph of the same location as in a) showing boundaries with large misorientation (larger than 30°C) between the adjacent grains with black lines.

3.2 Early phase of evolution: Effects of recovery and coarsening of the microstructures

Figure 7 shows a collage of micrographs that are all taken from the same solder interconnection. It should be noticed that we are using the same interconnection to exemplify the features of microstructural evolution that take place consequently in the stress concentration regions of the solder interconnections. The evolution of microstructures on the PWB side interfacial region is much slower than the evolution on the package side of the interconnection as a result of the less extensive plastic deformation per cycle (see Fig. 5). Even though this interconnection has experienced more than 3000 cycles and failed from the package side interfacial region, the microstructures visible on the PWB side of the interconnections are very similar to those on the package side interfacial regions about 1000 to 1500 cycles earlier.

As already discussed, the as-solidified microstructures of tin-rich solder interconnections are typically composed of relatively few large tin colonies distinguished by high-angle boundaries. The cellular structure of the tin-rich colonies is clearly visible within the high-angle boundaries as the individual tin cells are surrounded by eutectic regions composed of fine Cu_6Sn_5 and Ag_3Sn particles. As shown in Figure 7b the cellular structure of the primary tin-rich phase is still visible in the less deformed regions on the solder interconnections, such as in the middle, even after the component board has been thermally cycled until failure. However, in the strain concentration regions the tin cells begin to rearrange by the gradual coalescence of the tin cells and coarsening of the intermetallic particles (see Fig. 7c-d), during which the eutectic structures around the tin cells gradually disappear and a network

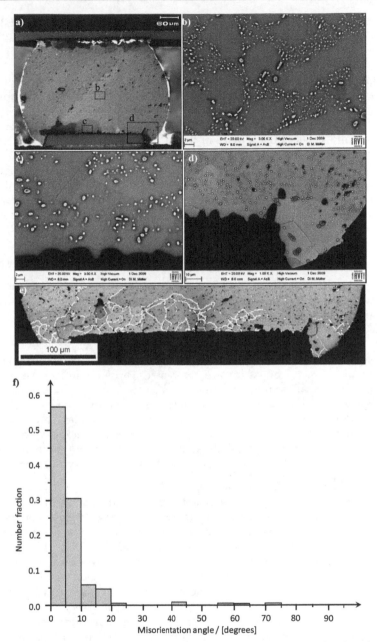

Fig. 7. a) Cross-polarized light micrograph of a solder interconnection that has failed under thermal cycling; b-d) magnifications from the regions indicated in (a); e) EBSD map of the cross-section that shows boundaries with a misorientation < 15° by yellow lines, 15°-45° by green lines, and > 45° by red lines; f) histogram distribution of grain boundaries over the region shown in (e). [64,65]

of low-angle boundaries produced by recovery emerges. Figure 7e shows an EBSD map of the PWB side interfacial region of the same cross-section as shown in Figure 7a. The yellow lines represent the low-angle boundaries, i.e., boundaries where the crystal orientation between the adjacent grains is *well* below 15° (see also Fig. 7f). The green and red lines represent high-angle boundaries where the misorientation between adjacent grains is higher than 15°, the green line those between 15° and 45°, and the red line those above 45°. The figure shows that in addition to the coarsening of the intermetallic particles, the formation of additional low-angle boundaries (additional to those between the cells of the Sn-rich phase formed during solidification) takes place in the regions of high strain energy density. These changes were observed to initiate early in the course of the thermal cycling tests, within about 50 to few hundred thermal cycles. It is also noteworthy that the coarsening of the intermetallic particles is strong in the regions near the high-angle grain boundaries, while the regions near the low-angle boundaries still include finer particles, comparable to the bulk of the solder ball (see Fig. 7d-e). This can be expected as the diffusion is much faster along the high-angle boundaries than it is along the low-angle boundaries.

3.3 Later phase of evolution: Transformation of the microstructure by recrystallization

In the course of further thermal cycling, the microstructures keep evolving gradually in the manner already described until these regions change discontinuous into more or less equiaxed grain structures by recrystallization. This change in the microstructures is first observed in the edge regions (i.e. corner regions of a cross-section) of the solder interconnections on the package side of the interconnections (the regions with the highest strain; see Fig.6) and, after the initiation of recrystallization, the recrystallized volume gradually expands from the edges toward the center, across the interconnections near the package side interfacial region of the interconnections. The incubation time of recrystallization varies significantly from one interconnection to another and even from one package to another (the same location of the interconnection) under the same loading conditions. The first indications of recrystallization in the BGA packaged board assemblies shown in Figure 1 were observed after about 500 thermal cycles but it can take up to about 2000 cycles until recrystallization is consistently observed in every interconnection in the corner regions of the packages.

Figure 8 shows a typical example of a failed interconnection, where the cracking of the solder interconnection is accompanied by a distinct change in the microstructure by recrystallization. The micrograph in Figure 8a is an optical micrograph that shows the crack path distinctly and the micrograph in Figure 8b is a cross-polarized light image of the same location as Figure 8a. The comparison of the micrographs shows that the propagation path of the crack is enclosed entirely within the recrystallized region of the interconnection. Figure 8c shows an EBSD map of the same cross-section as shown in Figures 8a-b. The colour lines represent the misorientation between the adjoining regions: yellow below 15°, green between 15° and 45°, and red above 45°. The image illustrates well the fact that the cracked region on the package side interface of the solder interconnections shows primarily high misorientations and that the high-angle grain boundaries are located very close to the crack path while, in general, the misorientations become smaller with increasing distance from the crack region. Figure 8d shows a histogram distribution of grain boundaries with different orientation over the region shown in Figure 8c (see also Fig 7f). As can be seen, the region still contains a high number of low-angle boundaries (caused by recovery) but a large

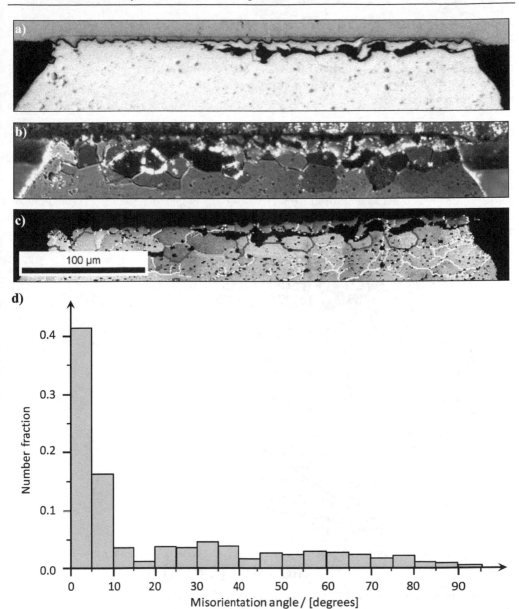

Fig. 8. a) an optical micrograph of a failed solder interconnection shows the crack path clearly (the thin black line between the gray intermetallic layer and the white solder is a contrast effect caused by specimen preparation); b) a cross-polarized light image of the same location as (a) highlights the recrystallized grains as caused by the cyclic deformation; c) EBSD map of the cross-section that shows boundaries with misorientation < 15° by yellow lines, 15°-45° by green lines, and > 45° by red lines; d) histogram distribution of grain boundaries over the region shown in (c). [64,65]

number of higher-angle boundaries (caused by recrystallization) have emerged. After the initiation of recrystallization in the strain concentration regions at the edges of the solder interconnections, the recrystallized volume gradually expands over the diameter of the interconnections, and cracks follow the expansion of the microstructurally changed volume. It was also observed that cracks rarely propagate outside the recrystallized volume of the solder interconnections.

4. Cracking of recrystallized solder interconnections

Work presented in more detail elsewhere focused on the evaluation of the nucleation time and propagation rate of cracks in BGA component board assemblies under different thermal cycling conditions [74,75]. Figure 9 shows the average crack lengths of the most critical solder interconnections as measured from cross-sections prepared along the diagonal line of the package as a function of thermal cycles. The different lines represent different cycling conditions (TS = thermal shock, TC = thermal cycling; the accompanying value is the dwell time of the profile in minutes).

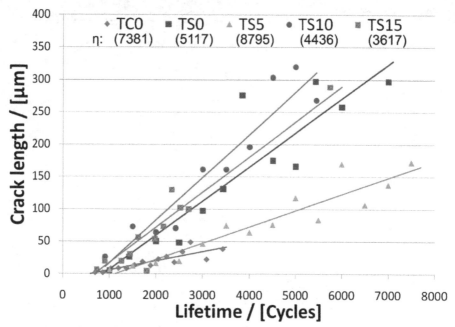

Fig. 9. Measured average crack lengths as a function of the number of thermal cycles (the numbers in parentheses are estimates of the Weibull characteristic lifetimes [η]). [74,75]

More detailed examinations showed that the nucleation of cracks in the interconnections of the BGA board assemblies took place within a relatively narrow range, between about 1000 and 1500 cycles, regardless of the dwell time or ramp rate used in the thermal cycling tests. However, the propagation rate of cracks without the influence of recrystallization was very slow. This conclusion was made based on the comparison of the measured crack lengths in interconnections that were removed from the thermal cycling oven at the same time and that showed or did not show recrystallization. Thus, the primary failure mechanism under thermomechanical fatigue involves the formation of a continuous network of grain boundaries by recrystallization that enables cracks to nucleate and propagate intergranularly through the solder interconnections (see Fig. 10).

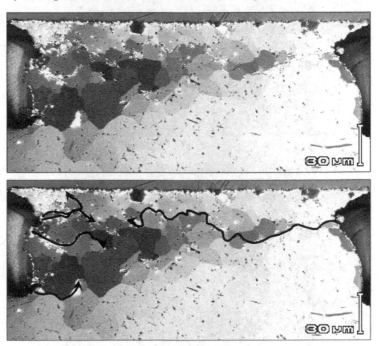

Fig. 10. Cracks propagate intergranularly between the recrystallized grains: a) cross-section of a recrystallized and cracked solder interconnection; b) the same image as in (a) but with superimposed crack paths.

The fractographic examinations illustrate the influence of recrystallization on crack propagation. Figure 11 shows the fractographs of failed solder interconnections. The fracture surfaces exhibit a globular appearance as a result of the propagation of cracks between the recrystallized grains. Fatigue striations were occasionally observed on the inspected fracture surfaces. Although they were quite uncommon, they indicate that cracks can also propagate transgranularly under conditions when the intergranular propagation is not benignant, namely when the recrystallized grain size is larger or the stress state, orientation, and geometry of the grains are not in favor of cracking along the grain boundaries. The networks of the grain boundaries formed by recrystallization evidently provide favorable paths for cracks to propagate intergranularly with less energy consumption in comparison with transgranular propagation. It can also be expected that the cohesion between the recrystallized tin grains is lowered by grain segregation of impurities as well as locally by intermetallic particles. Furthermore, the mechanical anisotropy of the (recrystallized) tin grains can also enhance the nucleation and propagation of microcracks along their boundaries as the value of the coefficient of thermal expansion of tin single-crystal in the [100] = [010] directions is about two times that in the [001] direction [50,51].

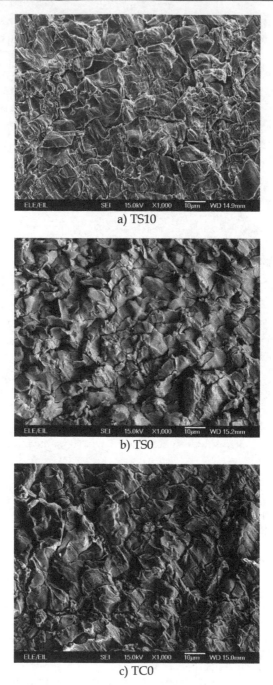

a) TS10

b) TS0

c) TC0

Fig. 11. Fractographs of the interconnections viewed from the component side [75]. Note the several secondary cracks visible in all micrographs (striations are not visible here).

As discussed in the previous chapter, the range of the incubation periods of recrystallization in tin-rich solder interconnections is relatively large, in the range of 500 to 2000 cycles in the case of the BGA interconnections in our study. Thus, the nucleation of cracks can take place before or after the change in microstructures by recrystallization. However, the failure analyses showed that the formation of the networks of grain boundaries by recrystallization had influenced the propagation of cracks in all electrically failed interconnections. The range of 1000-1500 thermal cycles required to produce distinguishable small cracks (nuclei) equals about 25-30% of the average lifetime of the component board assemblies and, thus, cracks are in the propagation stage for about three quarters of the lifetime of component boards. Therefore, it is evident that the cracking of tin-rich solder interconnections is controlled by the rate of recrystallization.

On the basis of the results presented above we can draw two conclusions: 1) the nucleation of cracks in solder interconnections is primarily dependent on the number of load reversals. In other words, nucleation is relatively insensitive of microstructural features and their evolution, as well as the parameters of the loading conditions; 2) the rate of crack propagation is dependent on the expansion rate of the recrystallized volume. In other words, the lifetimes of solder interconnections are primarily controlled by the onset and expansion of recrystallization.

5. An approach to lifetime prediction based on the evolution of microstructures

A thorough understanding of the restoration process in solders can allow the development of methods for improved lifetime estimation that are based on the evolution of microstructures. Work presented in [76] describes an approach to lifetime prediction based on the competing nature of the restoration processes: under conditions in which the strain hardening is more effective than recovery, the cyclic deformation accumulates the stored energy above a critical value, after which the recrystallization can initiate. The total stored energy of the system consists of the grain boundary energy and the volume defect energy (mainly line and point defects). The stored energy is released through the nucleation and growth of new strain-free grains (grains with low defect density), which gradually consume the strain-hardened matrix of high defect density. Li et al. have developed a multiscale model based on this principle for predicting the microstructural changes of recovery, recrystallization, and grain growth in solder interconnections subjected to dynamic loading conditions [77,78]. The approach developed in this work is based on the principle that the stored energy of the solder is gradually increased during each thermal cycle. When a critical value of the energy is reached, recrystallization is initiated. It is assumed that, even though recovery consumes a certain amount of the energy, the net change in the energy per cycle is always positive as experimental investigations have shown that newly recrystallized grains consistently appear sooner or later under various thermal cycling conditions. The stored energy is released through the nucleation and growth of new grains, which gradually consume the strain-hardened matrix of high defect density.

The approach is realized by combining Monte Carlo simulations with finite element calculations. The Monte Carlo method is employed to model the mesoscale microstructure and the finite element method to model the macroscale non-homogeneous deformation (see

Fig. 12). The non-homogeneous volume stored energy distribution in solder interconnections is scaled from the finite element model results and mapped onto the lattice of the Monte Carlo model. The quantitative prediction of the onset of recrystallization is carried out with the help of the Monte Carlo simulation.

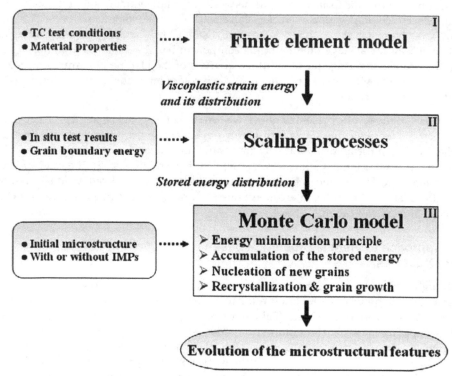

Fig. 12. Flow chart for the simulation of microstructural changes in solder interconnections [77,78]. Acronym TC stands for 'thermal cycling'.

In the Monte Carlo lattice, two adjacent sites with different grain orientation numbers are regarded as being separated by a grain boundary, while a group of sites with the same orientation number are considered a single grain. The total stored energy of the system under consideration consists of the grain boundary energy and the volume defect energy. Each site contributes an amount of stored energy to the system, and each pair of dissimilarly oriented neighboring sites contributes a unit of grain boundary energy to the system. The recrystallization process is modeled by randomly introducing nuclei (small embryos with zero stored energy) into the Monte Carlo lattice at a constant rate. An non-recrystallized site will become recrystallized if (a) the volume stored energy of the chosen sites is larger than the critical stored energy, and (b) the total energy of the system is reduced. If the selected site is recrystallized, it is considered as a contribution to the grain growth process.

The computational results were compared with the experimentally observed microstructural changes in solder interconnections subjected to thermal cycling tests. The results of the microstructural simulations carried out in this work can be summarized as follows: the incubation period of the recrystallization is about 1000 thermal cycles under the particular cycling conditions. The recrystallization is initiated in the corner regions on the package side of the solder interconnections and the expansion of the recrystallized region is controlled by the volume stored energy distribution. The expansion takes place first along the package side interfacial region toward the center of the interconnections. After that the recrystallized region expands toward the rest of the interconnections.

The incubation period of the recrystallization, the expansion of the recrystallized region, and the rate of increase of the recrystallized fraction are in good agreement with the experimental observations of thermally cycled component boards. This method predicts reasonably well the incubation period and the growth rate of the recrystallization, as well as the expansion of the recrystallized region. Figure 13 shows a comparison of predicted microstructural evolution with experimental evolution. The onset of recrystallization is a useful criterion to determine when the material models for the as-solidified microstructures are not valid anymore and, therefore, crack nucleation and propagation should be taken into account.

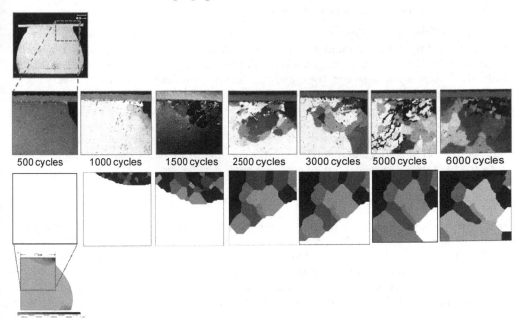

| 500 cycles | 1000 cycles | 1500 cycles | 2500 cycles | 3000 cycles | 5000 cycles | 6000 cycles |

Fig. 13. Observed and simulated microstructural changes of solder interconnections with an increasing number of thermal cycles [77].

In a more recent work Li et al. [78] have developed the method to take into consideration the fact that the nucleation of new grains of low defect density is more likely at the boundaries of high misorientation, such as the boundary between the intermetallic particles and the tin matrix. In particular, the coarse primary intermetallic particles can generate localized stress concentrations under an applied load because of their dissimilar mechanical properties with respect to the tin matrix. The intermetallic particles are introduced in the Monte Carlo model as inert particles that also do not move or grow. In practice the fine and uniformly distributed intermetallic particles would suppress the recrystallization to some extent by influencing the motion of the grain boundaries of recrystallizing grains but this is not (yet) considered in the model.

6. Conclusions

The cracking of the near-eutectic SnAgCu interconnections under thermal cycling conditions occurs through the bulk of the solder interconnections after a change in the microstructure by recrystallization. The cumulative increase in the stored energy during each deformation cycle provides the driving force for the recrystallization. After the solidification structure has changed into a more or less equiaxed grain structure, there is a continuous network of new high-angle boundaries providing favorable sites for cracks to nucleate and propagate with less energy consumption in comparison with the cracking of the as-soldered microstructure. The decrease in the stored energy in near-eutectic SnAgCu solders is assumed to take place very effectively by the recovery resulting from the high stacking fault energy of tin. Therefore, the recrystallization is initiated under well-defined loading conditions: dynamic loading where strain hardening is more effective than recovery.

The recrystallization of solder interconnections under thermomechanical (or in cyclic power) loading is an important phenomenon for the following reasons: being an experimentally observable indicator of microstructural evolution, the recrystallization enables one to establish a correlation between the field use loading conditions and those produced in accelerated reliability tests. Furthermore, the theoretically well-known phenomena of recovery and recrystallization can provide the means to incorporate the effects of microstructural evolution into lifetime prediction models, which are being increasingly employed to reduce the amount of reliability testing. Finally, by controlling the recrystallization of solder interconnections, for example by alloying, one may discover new solutions to improve the reliability of soldered electronic devices.

7. Acknowledgements

The authors wish to thank the following people for their valuable contribution to the work presented in this chapter: Mr. Jussi Hokka, Dr. Jue Li, Dr. Maik Mueller (TU Dresden, Germany), Mr. Otso Ratia, and Dr. Hongbo Xu. The authors would also like to thank the following people for their much appreciated help and the numerous discussions over the years: Ms. Pirjo Kontio, Ms. Sini Niiranen, Ms. Johanna Koivisto, Ms. Riitta Viitala, Dr. Hongtao Chen, Dr. Erkki Heikinheimo, Mr. Simo Miettinen, Dr. Tomi Laurila, and Dr. Vesa Vuorinen. Special thanks go to Prof. Mervi Paulasto-Kröckel and Prof. Klaus-Juergen Wolter (TU Dresden, Germany) for their favorable support for this work. The financial support from the Academy of Finland (decision number 123922), the Finnish Funding Agency for Technology and Innovation (decision numbers 40135/07, 662/06, and the ELMO program), and the Finnish electronics industry is gratefully acknowledged.

8. References

[1] J. Karppinen, T. T. Mattila, and J. K. Kivilahti, "Formation of thermomechanical interconnection stresses in a high-end portable product," *The Proceedings of the 2nd Electronics System Integration Technology Conference*, London, UK, September 1-4, 2008, IEEE/EIA CPMT, (2008), pp. 1327-1332.

[2] J. S. Karppinen, J. Li, T. T. Mattila, and M. Paulasto-Kröckel, "Thermomechanical reliability characterization of a handheld product in accelerated tests and use environment," *Microelectronics Reliability*, (in print).

[3] IEC 60068-2-14 Ed. 5.0 b: 1984, "Environmental testing – part 2: tests. Test N: change of temperature," International Electrotechnical Commission, (1984), 34 p.

[4] JESD22-A104C, "Temperature Cycling," Jedec Solid State Technology Association, (2005), 16 p.

[5] IPC-TM-650 rev. A, "Thermal Shock and Continuity, Printed Board," The Institute for Interconnecting and Packaging Electronic Circuits, (1997), 2 p.

[6] C. F. Coombs Jr., Printed Circuits Handbook, 5th ed., New York, (2001), McGraw-Hill, 1200 p.

[7] Y. S. Touloukian and C. Y. Ho, Thermal Expansion: Metallic Elements and Alloys, New York, (1975), IFI/Plenum, 316 p.

[8] Directive 2002/95/EC of the European Parliament and of the Council on the Restriction of the Use of Hazardous Substances in Electrical and Electronic Equipment (RoHS), Jan. 27th, 2003.

[9] Directive 2002/96/EC of the European Parliament and of the Council on Waste of Electrical and Electronic Equipment (WEEE), Jan. 27th, 2003.

[10] D. R. Frear, "Microstructural evolution during the thermomechanical fatigue of solder joints," in M.J. Cieslak, M.E. Glicksman, S. Kang, and M.E Glicksman, *The Metal Science of Joining*, The Minerals, Metals & Materials Society, pp. 191-2000.

[11] J. W. Morris Jr., D. Tribula, T. S. E. Summers, and D. Grivas, "The role of microstructure in thermal fatigue of Pb-Sn solder joints," in John H. Lau, *Solder Join Reliability*, New York, 1991, Van Nostrand Reinold, pp. 225-265.

[12] T. T. Mattila, T. Laurila, and J. K. Kivilahti, "Metallurgical factors behind the reliability of high density lead-free interconnections," in E. Suhir, C. P. Wong, and Y. C. Lee, *Micro-and Opto-Electronic Materials and Structures: Physics, Mechanics, Design, Reliability, Packaging*, Springer Publishing Company, New York, 2007, (1), pp. 313-350.

[13] W. Q. Peng, "Lead-free electronic assembly based on Sn–Ag–Cu solders," Espoo, licentiate thesis, Helsinki University of Technology, (2001), p. 124.

[14] K.-W. Moon, W. J. Boettinger, U. R. Kattner, F. S. Biancaniello, and C. A. Handwerker, "Experimental and thermodynamic assessment of Sn–Ag–Cu solder alloys," *Journal of Electronic Materials*, 29, 10, (2000), pp. 1122-1136.

[15] T. T. Mattila, V. Vuorinen, and J. K. Kivilahti, "Impact of printed wiring board coatings on the reliability of lead-free chip-scale package interconnections," *Journal of Materials Research*, 19,11, (2004), pp. 3214-3223.

[16] A. LaLonde, D. Emelander, J. Jeannette, C. Larson, W. Rietz, D. Swenson, and D. W. Henderson, "Quantitative metallography of β-Sn dendrites in Sn3.8Ag0.7Cu ball grid array solder balls via electron backscatter diffraction and polarized light microscopy," *Journal of Electronic Materials*, 33, 12, (2004), pp. 1545-1549.

[17] D. Henderson, J. J. Woods, T. A. Gosseling, J. Bartelo, D. E. King, T. M. Korhonen, M. A. Korhonen, L. P. Lehman, E. J. Cotts, S. K. Kang, P. Lauro, D.-Y. Shih, C. Goldsmith, and K. J. Puttliz, "The microstructure of Sn in near eutectic Sn-Ag-Cu alloy solder joints and its role in thermomechanical fatigue," *Journal of Materials Research*, 19, 6, (2004), pp. 1608-1612.

[18] S. Terashima and M. Tanaka, "Thermal fatigue properties of Sn-1.2Ag-0.5Cu-xNi Flip Chip interconnects," *Materials Transactions*, 45, 3, (2004), pp. 681-688.

[19] S. K. Kang, P. A. Lauro, D.-Y. Shih, D. W. Henderson, and K. J. Puttlitz, "Microstructure and mechanical properties of lead-free solders and solder joints used in microelectronic applications," *IBM Journal of Research and Development*, 49, 4/5, (2005), pp. 607-620.

[20] A. U. Telang, T. R. Bieler, J. P. Lucas, K. N. Subramanian, L. P. Lehman, Y. Xing, and E. J. Cotts, "Grain-boundary character and grain growth in bulk tin and bulk lead-free solder alloys," *Journal of Electronic Materials*, 33, 12, (2004), pp. 1412-1423.

[21] L. P. Lehman, S. N. Athavale, T. Z. Fullem, A. C. Giamis, R. K. Kinyanjui, M. Lowenstein, K. Mather, R. Patel, D. Rae, J. Wang, Y. Xing, L. Zavalij, P. Borgesen, and E. J. Cotts, "Growth of Sn and intermetallic compounds in Sn-Ag-Cu solder," *Journal of Electronic Materials*, 33, 12, (2004), pp. 1429-1439.

[22] Z. G. Chen, Y. W. Shi, Z. D. Xia, and Y. F. Yan, "Study on the microstructure of a novel lead-free solder alloy SnAgCu-RE and its soldered joints," *Journal of Electronic Materials*, 31, 10, (2002), pp. 1122-1128.

[23] Maik Müller, Steffen Wiese, and Klaus-Jürgen Wolter, "Influence of cooling rate and composition on the solidification of SnAgCu solders," *The Proceedings of the 1st Electronics Systemintegration Technology Conference*, Dresden, Germany, September 5 - 7, 2006, IEEE/EIA CPMT, (2006), pp. 1303-1311

[24] O. Fouassier, J.-M. Heintz, J. Chazelas, P.-M. Geffroy, and J.-F. Silvain, "Microstructural evolution and mechanical properties of SnAgCu alloys," Journal of Applied Physics, 100, 043519 (2006), pp. 043519-1 - 043519-8.

[25] J. Gong, C. Liu, P. P. Conway, and V. V. Silberschmidt, "Crystallographic structure and mechanical behaviour of SnAgCu solder interconnects under a constant loading rate," *The Proceedings of the 57th Electronic Component and Technology Conference*, Reno, NV, May 29- June 1, 2007, IEEE/EIA CPMT, (2007), pp. 677-683

[26] P. Lauro, S. K. Kang, W. K. Choi, and D.-Y. Shih, "Effects of mechanical deformation and annealing on the microstructure and hardness of Pd-free solders," *Journal of Electronic Materials*, 32, 12, (2003), pp. 1432-1440.

[27] M. Krause, M. Mueller, M. Petzold, S. Wiese, and K. J. Wolter, "Scaling effects on grain size and texture of lead free interconnects – Investigations by electron backscatter diffraction and nanointendation," *The Proceedings of the 58th Electronic Component and Technology Conference*, Orlando, FL, May 27-30, 2008, IEEE/EIA CPMT, (2008), pp. 75-81.

[28] T.-M. K. Korhonen, P. Turpeinen, L. P. Lehman, B. Bowman, G. H. Thiel, R. C. Parkes, M. A. Korhonen, D. W. Henderson, and K. J. Puttlitz, "Mechanical properties of near-eutectic Sn-Ag-Cu alloy over a wide range of temperatures and strain rates," *Journal of Electronic Materials*, 33, 12, (2004), pp. 1581-1588.

[29] W. Peng, K. Zeng, and J. Kivilahti, "A literature review on potential lead-free solder systems," Espoo, Helsinki University of Technology, Report Series HUT–EPT–1, (2000), 53 p.

[30] S. Chada, R. A. Fournelle, W. Laub, and D. Shangguan, "Copper substrate dissolution in eutectic Sn-Ag solder and its effect on microstructure," *Journal of Electronic Materials*, 29, 10, (2000), pp. 1214-1221.

[31] M. O. Alam, Y. C. Chan, and K. N. Tu, "Effect of 0.5 wt% Cu addition in Sn–3.5%Ag solder on the dissolution rate of Cu metallization," *Journal of Applied Physics*, 94, 12, (2003), pp. 7904-7909.

[32] W. G. Bader, "Dissolution of Au, Ag, Pd, Pt, Cu and Ni in a Molten Sin-Lead Solder," *Welding Journal*, 48, 12, (1969), pp. 551s-557s.

[33] W. G. Bader, "Dissolution and formation on intermetallics in the soldering process," *The proceedings of the Conference on Physical Metallurgy and Metal Joining*, St. Louis, MO, Oct. 16-17. 1980, Warrendale, USA.

[34] W. C. Leslie, T. J. Michalak, and F. W. Aul, "The annealing of cold-worked iron," in C. W. Spencer and F. E. Werner, *Iron and Its Dilute Solid Solutions*. New York, 1963, Interscience Puhlishers, pp. 103-119.

[35] R. W. Cahn, "Recovery and recrystallization," in R. W. Cahn, *Physical Metallurgy*, Amsterdam, 1965, North–Holland Publishing Company, pp. 925–987.

[36] F. J. Humphreys and M. Hatherly, *Recrystallization and Related Annealing Phenomena*, 2nd ed., Oxford, 2004, Elsevier Ltd., 574 p.

[37] R.D. Doherty, D. A. Hughes, F. J. Humphreys, J. J. Jonas, D. Juul Jensen, M. E. Kassner, W. E. King, T. R. McNelley, H. J. McQueen, and A. D. Rollett, "Current issues in recrystallization," *Materials Science and Engineering A*, 238, (1997), pp. 219 – 274.

[38] H. Yu and J. K. Kivilahti, "Nucleation kinetics and solidification temperatures of SnAgCu interconnections during reflow process," IEEE Transactions on Components and Packaging Technologies, 29, 4, (2006), pp. 778 - 786.

[39] R. Darveaux, C. Reichman, and P. Agrawal, "Solidification behavior of lead free and tin lead solder bumps," *The Proceedings of the 60th Electronic Component and Technology Conference*, Las Vegas, NV, June 1-4, 2010, IEEE/EIA CPMT, (2010), pp. 1442-1447.

[40] J. S. Kang, R. A. Gagliano, G. Ghosh, and M. E. Fine, "Isothermal solidification of Cu/Sn diffusion couples to form thin-solder joints," *Journal of Electronic Materials*, 31, 11, (2002), pp. 1238-1243.

[41] J.-M. Song, J.-J. Lin, C.-F. Huang, and H.-Y. Chuang, "Crystallization, morphology and distribution of Ag_3Sn in Sn–Ag–Cu alloys and their influence on the vibration fracture properties," *Materials Science and Engineering A*, 466, (2007), pp. 9-17

[42] D. W. Henderson, T. Gosselin, A. Sarkhel, S. K. Kang, W.-K. Choi, D.-Y. Shih, C. Goldsmith, and K. J. Puttlitz, "Ag_3Sn plate formation in the solidification of near ternary eutectic Sn–Ag–Cu alloys," *Journal of Materials Research*, 17, (2002), pp. 2775-2778.

[43] J. H. Perepezko, D. H. Rasmussen, I. E. Anderson, and C. R. Loper, Jr., "Undercooling of low-melting point metals and alloys," The *Proceedings of the International Conference on Solidification and Casting of Metals*, Sheffield, England, July 1977, Sheffield Metallurgical and Engineering Association / University of Sheffield / the Metals Society, (1979), pp. 169-174.

[44] S. Wiese, E. Meusel, and K. J. Wolter, "Microstructural dependence of constitutive properties of eutectic SnAg and SnAgCu solders," *The Proceedings of the 53rd Electronic Components and Technology Conference*, 27 May-30 May, 2003, New Orleans, LA, IEEE EIA/CPMT, (2003), pp. 197-206.

[45] S. K. Kang, W. K. Clioi, D.-Y. Shih, D. W. Henderson, T. Gossefin, A. Sarkliel, C. Goldsmith, and K. J. Puttlitz, "Formation of Ag_3Sn plates in Sn-Ag-Cu alloys and optimization of their alloy composition," *The Proceedings of the 53rd Electronic Components and Technology Conference*, 27 May-30 May, 2003, New Orleans, LA, IEEE EIA/CPMT, (2003), pp. 64-70.

[46] D. Swenson, "The effects of suppressed beta tin nucleation on the microstructural evolution of lead-free solder joints," *Journal of Material Science: Materials in Electronics*, 18, 1-3, (2007), pp. 39-54.

[47] S. K. Kang, G. C. Moon; P. Lauro, and D.-Y. Shih, "Critical Factors Affecting the Undercooling of Pb-free, Flip-Chip Solder Bumps and In-situ Observation of

Solidification Process," *The Proceedings of the 57th Electronic Component and Technology Conference*, Reno, NV, May 29-June 1, 2007, IEEE/EIA CPMT, (2007), pp. 1597-1603.

[48] J. W. Elmer, E. D. Specht, and M. Kumar, "Microstructure and in situ observations of undercooling for nucleation of β-Sn relevant to lead-free solder alloys," *Journal of Electronic Materials*, 39, 3, (2010), pp. 273-282.

[49] M. A. Matin, W. P. Vellinga, and M. G. D. Geers, "Thermomechanical fatigue damage evolution in SAC solder joints," *Materials Science and Engineering A*, 445-446, (2007), pp. 73-85.

[50] N. S. Brar and W. R. Tyson, "Elastic and plastic anisotropy of white tin," *Canadian Journal of Physics*, 50, 19, (1972), pp. 2257-2264.

[51] *Metals Handbook*, Volume 1 – Properties and Selection of Metals, 8th ed., New York, 1961, American Society for Metals, 1300 p.

[52] K. N. Subramanian, "Role of anisotropic behavior of Sn on thermomechanical fatique and fracture of Sn-based solder joints under thermal excursions," *Fatigue and Fracture of Engineering Materials and Structures*, 30, 5, (2007), pp. 420-431.

[53] M. A. Matin, E. W. C. Coenen, W. P. Vellinga, and M. G. D. Geers, "Correlation between thermal fatigue and thermal anisotropy in a Pb-free solder alloy," *Scripta Materialia*, 53, (2005), pp. 927-932.

[54] T. R. Bieler, H. Jiang, L. P. Lehman, T. Kirkpatrick, E. J. Cotts, and B. Nandagopal, "Influence of Sn grain size and orientation on the thermomechanical response and reliability of Pb-free solder joints," *IEEE Transactions on Components and Packaging Technologies*, 31, 2, (2008), pp. 370-380.

[55] T. R. Bieler, B. Zhou, L. Blair, A. Zamiri, P. Darbandi, F. Pourboghrat, T.-K. Lee, and K.-C. Liu, "The role of elastic and plastic anisotropy of Sn on microstructure and damage evolution in lead-free solder joints," *The Proceedings of the 2011 IEEE International Reliability Physics Symposium*, 10-14 April, 2011, Monterey, CA, IEEE, (2011), pp. 5F.1.1-5F.1.9.

[56] S. Terashima, K. Takahama, M. Nozaki, and M. Tanaka, "Recrystallization of Sn grains due to thermal strain in Sn-1.2Ag-0.5Cu-0.05N solder," *Materials Transactions, Japan Institute of Metals*, 45, 4, (2004), pp. 1383-1390.

[57] S. Dunford, S. Canumalla, and P. Viswanadham, "Intermetallic morphology and damage evolution under thermomechanical fatigue of lead (Pb)-free solder interconnections," The Proceedings of the 54th Electronic Components and Technology Conference, June 1-4, 2004, Las Vegas, NV, USA, IEEE/EIA/CPMT, (2004), pp. 726-736.

[58] L. Lehman, S. Athavale, T. Fullem, A. Giamis, R. Kinyanjui, M. Lowenstein, K. Mather, R. Patel, D. Rae, J. Wang, Y. Xing, L. Zavalij, P. Borgesen, and E. Cotts, "Growth of Sn and intermetallic compounds in Sn-Ag-Cu solder," *Journal of Electronic Materials*, 3, 12, (2004), pp. 1429-1439.

[59] P. Limaye, B. Vandevelde, D. Vandepitte, and B. Verlinden, "Crack growth rate measurement and analysis for WLCSP Sn-Ag-Cu solder joints," The proceedings of the SMTA international annual conference, Chicago, IL, September 25-29, (2005), pp. 371-377.

[60] L. Xu and J. H.L. Pang, "Intermetallic growth studies on SAC/ENIG and SAC/CU-OSP lead-free solder joints," *Thermal and Thermomechanical Phenomena in Electronics Systems*, (2006), pp.1131-1136.

[61] A. U. Telang, T. R. Bieler, A. Zamirini, and F. Pourboghrat, "Incremental recrystallization/crain growth driven by elastic strain energy release in a thermomechanically fatigued lead-free solder joint," *Acta Materialia*, 55, (2007), pp. 2265-2277.

[62] J. J. Sundelin, S. T. Nurmi, and T. K. Lepistö, "Recrystallization behavior of SnAgCu solder joints," *Materials Science and Engineering A*, 474, (2008), pp. 201-207.

[63] J. Li, J. Karppinen, T. Laurila, and J. K. Kivilahti, "Reliability of Lead-Free solder interconnections in Thermal and Power cycling tests," *IEEE Transactions on Components and Packaging Technologies*, 32, (2009), pp. 302-308.

[64] H. T. Chen, M. Mueller, T. T. Mattila, J. Li, X.W. Liu, K.-J. Wolter, and M. Paulasto-Kröckel, "Localized Recrystallization and Cracking of Lead-Free Solder Interconnections under Thermal Cycling," *Journal of Materials Research*, 25, 16, (2011), pp. 2103-2116.

[65] T. T. Mattila, M. Mueller, M. Paulasto-Kröckel, and K. J. Wolter "Failure mechanism of solder interconnections under thermal cycling conditions," *The Proceedings of the 3rd Electronic System-Integration and Technology Conference*, Berlin, Germany, September 13-16, 2010, IEEE/EIA CPMT, (2010), pp. 1-8.

[66] B. Zhou, T. T. Bieler, T. K. Lee, and K.C. Liu, "Crack development in a low-stress PBGA package due to continuous recrystallization leading to formation of orientations with [001] parallel to the interface," *Journal of Electronic Materials*, 39, (2010), pp. 2669 -2679.

[67] D. Hardwick, C. M. Sellars, and W. J. McG. Tegart, "The occurrence of recrystallization during high-temperature creep," *Journal of the Institute of Metals*, 90, (1961), pp. 21-22.

[68] D. McLean and M. H. Farmer, "The relation during creep between grain–boundary sliding, sub-crystal size, and extension," *Journal of Institute of Metals*, 85, (1956), pp. 41-50.

[69] P. Gay and A. Kelly, "X-ray studies of polycrystalline metals deformed by rolling. II. Examination of the softer metals, tin, zinc, lead and cadmium," *Acta Crystallographica*, 6, (1953), pp. 172-177.

[70] A. G. Guy, *Elements of Physical Metallurgy*, 2nd ed., London, 1960, Addison-Wesley Publishing Company Inc., 528 p.

[71] S. Miettinen, *Recrystallization of Lead-free Solder Joints under Mechanical Load*, Master's Thesis (in Finnish), Espoo, (2005), 84 p.

[72] T. M. Korhonen, L. Lehman, M. Korhonen, and D. Henderson, "Isothermal fatigue behavior of the near-eutectic Sn-Ag-Cu alloy between -25°C and 125°C," *Journal of Electronic Materials*, 36, 2, (2007), pp. 173-178.

[73] T. Laurila, T. T. Mattila, V. Vuorinen, J. Karppinen, J. Li, M. Sippola, and J. K. Kivilahti, "Evolution of microstructure and failure mechanism of lead-free solder interconnections in power cycling and thermal shock tests," *Microelectronics Reliability*, 47, 7, (2007), pp. 1135-44.

[74] T. T. Mattila, H. Xu, O. Ratia, and M. Paulasto-Kröckel, "Effects of thermal cycling parameters on lifetimes and failure mechanism of solder interconnections," *The Proceedings of the 60th Electronic Component and Technology Conference*, Las Vegas, NV, June 1-4, 2010, IEEE/EIA CPMT, (2010), pp. 581-590.

[75] H. Xu, T. T. Mattila, O. Ratia, and M. Paulasto-Kröckel, "Effects of thermal cycling parameters on lifetimes and failure mechanism of solder interconnections," *IEEE Transactions on Manufacturing and Packaging Technologies – a Special Issue*, (in print). Invited paper.

[76] T. T. Mattila and J. K. Kivilahti, "The role of recrystallization in the failure mechanism of SnAgCu solder interconnections under thermomechanical loading," *IEEE Transactions on Components and Packaging Technologies*, 33, 3, (2010), pp. 629-635.

[77] J. Li, T. T. Mattila, and J. K. Kivilahti, "Multiscale simulation of recrystallization and grain growth of Sn in lead-free solder interconnections," *Journal of Electronic Materials*, 39, 1, (2010), pp. 77-84.

[78] J. Li, H. Xu, T. T. Mattila, J. K. Kivilahti, T. Laurila, and M. Paulasto-Kröckel, "Simulation of dynamic recrystallization in solder interconnections during thermal cycling," *Computational Materials Science*, 50, (2010), pp. 690-697.

7

Phase Transformations and Recrystallization Processes During Synthesis, Processing and Service of TiAl Alloys

Fritz Appel

Institute for Materials Research, Helmholtz-Zentrum Geesthacht, Geesthacht, Germany

1. Introduction

Titanium aluminides alloys based on the intermetallic phases $\alpha_2(Ti_3Al)$ and $\gamma(TiAl)$ are one of the few classes of emerging materials that have the potential for innovative applications in advanced energy conversion systems whenever low density, good high-temperature strength and resistance against ignition and corrosion are of major concern [1]. The outstanding thermo-physical properties of the individual phases mainly result from the highly ordered nature and directional bonding of the compounds. However, two-phase $\alpha_2(Ti_3Al)+\gamma(TiAl)$ alloys exhibit a much better mechanical performance than their monolithic constituents $\gamma(TiAl)$ and $\alpha_2(Ti_3Al)$, provided that the phase distribution and grain size are suitably controlled. The synergistic effects of the two phases are undoubtedly associated with the many influences that the microstructure has on deformation and fracture processes. Constitution and microstructure are the result of phase transformations, ordering reactions and recrystallization processes, which occur during synthesis, processing and service. Many aspects of these mechanisms are intimately linked to defect configurations at the atomic level; thus standard techniques of metallography were often inadequate to provide the necessary information. This lack of information is addressed in the present article in that observations on recrystallization and phase transformations by high-resolution electron microscopy are presented. Particular emphasis will be paid on

i.   the origin of microstructures
ii. heterogeneities in the deformed state and recovery behaviour
iii. atomic structure of crystalline and crystalline/amorphous interfaces
iv. misfit accommodation and coherency stresses.

2. Constitution and microstructure

2.1 Constitution

TiAl alloys of technical significance have the general composition (in at. %, as are all compositions in this paper)

$$Ti\text{-}(42\text{-}49)Al+X, \tag{1}$$

with X designating alloying elements, such as Cr, Nb, W, V, Ta, Si, B, and C [2]. When referred to the binary phase diagram (Fig. 1), the equilibrium phases for Al contents between 46 and 49 % are: the disordered solution phases hexagonal (h.c.p.) α(Ti), body centred cubic (b.c.c.) β(Ti), and the ordered intermetallic compounds γ(TiAl) with $L1_0$ structure, and α_2(Ti$_3$Al) with DO_{19} structure [3]. Based on this constitution, different microstructures have been designed. The phase transformations involved in the microstructural evolution will be demonstrated for two examples, the classical lamellar structure of α_2(Ti$_3$Al)+γ(TiAl) alloys and a novel crystallographically modulated morphology occurring in multiphase alloys.

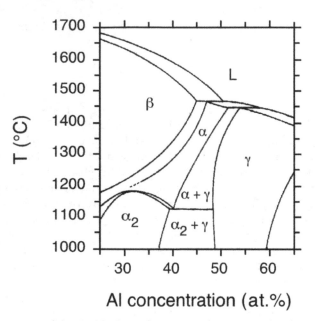

Fig. 1. Central portion of the Ti-Al-phase diagram in the region of technical interest [3].

2.2 Lamellar microstructure

The technologically most relevant α_2(Ti$_3$Al)+γ(TiAl) alloys contain a significant volume fraction of lamellar grains [1, 2]. The morphology of these grains represents a multilayer system made of two phases. It is well documented in the literature [4] that such a system could exhibit extraordinary mechanical properties when the layer thickness is small enough.

The so-called lamellar microstructure results from the precipitation of γ lamellae in either a disordered α or a congruently ordered α_2' matrix, following one of the transformation paths [3]:

$$\alpha \rightarrow \alpha_2' \rightarrow \alpha_2 + \gamma \text{ or } \alpha \rightarrow \alpha_2 + \gamma. \tag{2}$$

α_2' and α_2 have the same crystal structure but different composition. The exact pathway is still a matter of debate and could depend on alloy composition and thermal treatment. The orientation relationships between the α_2 and γ platelets are [5]

$$\{111\}\gamma \ || \ (0001)\alpha2 \ \text{and} \ [1\,\overline{1}\,0]\gamma \ || \ <11\,\overline{2}\,0>\alpha2. \tag{3}$$

The length of the lamellae is determined by the size of the parent α/α_2 grain. The γ phase is formed as an ordered domain structure, as sketched in Fig. 2. This gives rise to six variants of the above orientation relationships. Thus, there are four types of lamellar interfaces: the α_2/γ interface and three distinct γ/γ interfaces that are typified by rotations of 60°, 120° and 180° between adjacent lamellae. A lamellar grain consists of a set of γ lamellae, which are subdivided into domains and interspersed by α_2 lamellae. The volume fraction of the two phases is controlled by the composition on the basis of the phase diagram and the processing conditions of the alloy. However, it should be noted that the decomposition of α phase into lamellar $(\alpha_2+\gamma)$ is sluggish and often cannot be established within the constraints of processing routes; thus the volume fraction of γ phase is less than equilibrium [6]. It might be expected that the Ti concentration in the γ phase increases when the alloy becomes richer in Ti until the maximum solubility of Ti in the γ phase is reached. This non-equilibrium phase composition may provide significant driving forces for structural changes, as will be outlined in the subsequent sections.

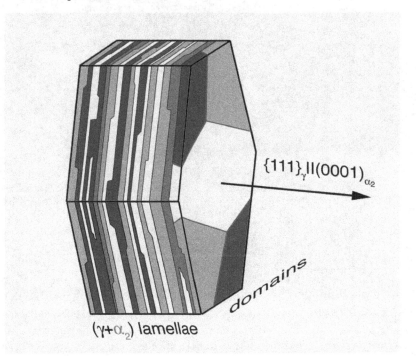

$\{111\}_\gamma || (0001)_{\alpha2}$

domains

$(\gamma+\alpha_2)$ lamellae

Fig. 2. Schematic drawing of a lamellar colony comprised of γ and α_2 platelets.

Among the various lamellar interfaces only the twin boundary is fully coherent, as the adjacent lattices are symmetrically oriented. At all the other interfaces the matching is imperfect, i.e., these interfaces are semicoherent. The mismatch arises from the differences in the crystal structure and lattice parameters and amounts to 1 to 2 %, depending on alloy composition and processing conditions. Different modes of mismatch accommodation have been discussed for lamellar TiAl alloys, which in broad terms correspond to the early models of misfitting interfaces [7]. Up to a certain point, the misfit strain could be solely taken up by elastic distortion, i.e., the lamellae are uniformly strained to bring the atomic spacings into registry. This homogeneous strain accommodation leads to coherent interfaces but introduces lattice distortions that are known as coherency strains. Hazzledine [4] has shown that the elastic misfit accommodation in lamellar $(\alpha_2+\gamma)$ alloys is only possible if the lamellae are very thin. The predicted critical thicknesses are $d_c \leq 8$ nm for the mismatched γ/γ interfaces and $d_c \leq 0.8$ to 3.9 nm for the α_2/γ interfaces, depending on the volume content of α_2 phase. Figure 3 shows a small Ti_3Al platelet embedded in γ phase with a thickness of 4.5 nm, which is just above the coherency limit predicted by Hazzledine [4]. While homogeneous strain accommodation is still recognizable, part of the misfit is already taken up by interfacial defects.

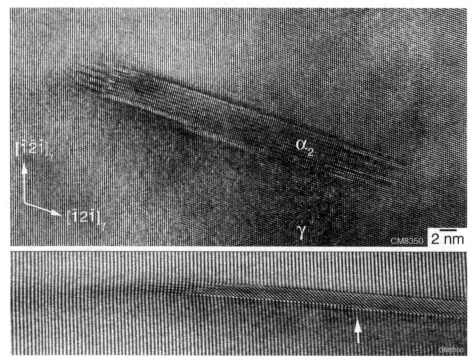

Fig. 3. High-resolution electron micrograph showing a small Ti_3Al platelet embedded in γ phase. Interfacial steps, dislocations and homogeneous elastic straining accommodate the misfit between the particle and the matrix. In the compressed image below, the elastic straining at the tip of the particle is readily visible by the distortion of the $(\bar{1}1\bar{1})$ planes. A dislocation compensating the misfit between the $(\bar{1}1\bar{1})_\gamma$ and $\{2\bar{2}01\}_{\alpha2}$ planes is arrowed. Ti-46.5Al-4(Cr, Ta, Mo, B), sheet material.

The coherency strains raise the total energy of the system. Thus, for a sufficiently large misfit, or lamellar spacing, it becomes energetically more favourable to replace the coherent interface by a semicoherent interface, a situation that is referred to as a loss of coherency. As described in the early model of Frank and van der Merve [8] misfit dislocations partially take up the misfit, i.e., the atoms at the interface adjust their positions to give regions of good and bad registry. In other words, the misfit is concentrated at the dislocations.

Mismatch strains at interfaces can be also relieved by the formation of ledges or steps at the interface [9]. The introduction of mono-atomic steps significantly improve the atomic matching, thus preventing the disregistry from becoming large anywhere. Structural ledges may replace misfit dislocations as a way of retaining low-index terraces between the respective defects. The general view is that planar boundaries are favoured for large misfits and small Burgers vectors of the misfit dislocations, whereas stepped boundaries are favoured for small misfits and large values of the Burgers vector. Steps may range in scale from atomic to multi-atomic dimensions depending on energetic or kinetic factors. However, it is very often the case that an interfacial defect exhibits both dislocation- and step-like character, thus, comprising a more general defect that has been defined as a disconnection [10,11]. Because of its step character, disconnection motion along an interface transports material from one phase to the other, the extent of which is essentially determined by the step height. At the same time, the dislocation content of the disconnection leads to deformation. In this sense, disconnection motion couples interface migration with deformation. The different extents of symmetry breaking at the interfaces lead to a broad variety of disconnections [10, 11]. This is reflected in different step heights and dislocation contents, which eventually determine the function of disconnections in phase transformations. Disconnection models have been developed for a variety of diffusional and diffusionless phase transformations in crystalline solids; and an extensive body of literature has evolved. For more details the reader is referred to a review of Howe et al. [11]. Several authors [12-16] observed interfacial steps at α_2/γ interfaces with heights that were always a multiple of $\{111\}_\gamma$ planes. The most commonly observed two-plane steps were characterized as disconnections with the topological parameters $b=1/6[11\bar{2}]$, $t(\gamma)=1/2[\overline{112}]_\gamma$ and $t(\alpha_2)=[000\bar{1}]_{\alpha2}$ [17]. b is the Burgers vector of the dislocation component (parallel to the (111) interface) and $t(\gamma)$ and $t(\alpha_2)$ are vectors describing the ledge risers [11] of the disconnection. This type of disconnection has no Burgers vector component perpendicular to the interface.

The complexity of misfit accommodation at α_2/γ interfaces is illustrated in Fig. 4. The micrograph shows a γ(TiAl) lamella terminated within the α_2 phase of a two-phase alloy. The interface marked in the micrograph borders the crystal region in which the exact ABC stacking of the $L1_0$ structure is fulfilled. Outside of this exactly stacked region of γ phase, there is a two to three atomic plane thick layer in which neither the ABC stacking of the γ phase nor the $ABAB$ stacking of the α_2 phase is correctly fulfilled. This becomes particularly evident at the tip of the γ lamellae and indicates a significant homogeneous straining of the lattice. This strain seems to locally relax by the formation of dislocations, as can be seen in the compressed form of the image. The other salient feature is the misfit accommodation by steps. A Burgers circuit constructed around the tip of the γ lamella results in a projected Burgers vector of $b_p=1/6[11\bar{2}]$. This indicates that the small misfit between the $(111)_\gamma$ and $(0002)_{\alpha2}$ planes is elastically taken up. The observed misfit accommodation is pertinent to the issue of how differential material flux during the $\alpha_2 \rightarrow \gamma$ transformation is accomplished.

There is ample evidence [18] of enhanced self-diffusion along dislocation cores. Likewise interfacial ledges are envisaged as regions where deviation from the ideal structure is localized and which may provide paths of easy diffusion. This gives rise to the speculation that it is mainly the tip of a newly formed γ lamella where the atomic composition between the two phases is adjusted during transformation.

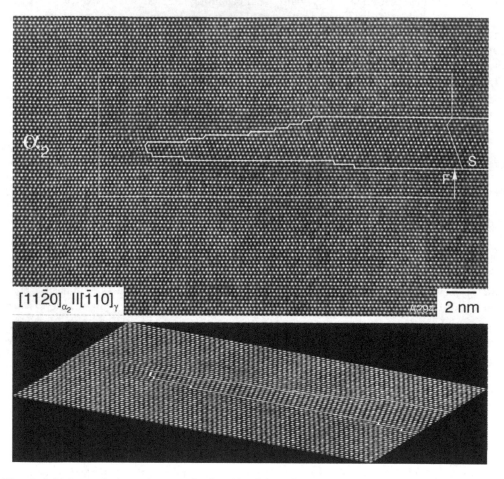

Fig. 4. A high-resolution micrograph of a γ(TiAl) lamella terminated within α_2 phase imaged down the $<10\bar{1}>_{\gamma}$ and $<11\bar{2}0>_{\alpha 2}$ directions. The thick white line marks the position of the interface. The salient feature is the misfit accommodation by steps. S and F denote the start and finish, respectively, of the Burgers circuit constructed around the tip of the γ lamella. After elimination of all the cancelling components in the initial circuit, and transforming the sequence of operations in the α_2 phase into the γ coordinate frame, the projected Burgers vector is $b_p = 1/6[11\bar{2}]$. Note the dislocation in front of the γ tip. This can be recognized in the image below, which was compressed along the $(2\bar{2}01)_{\alpha 2}$ planes. Ti-Al-Nb, as cast.

In spite of the misfit accommodation by interfacial dislocations and ledges a significant elastic strain remains at the interfaces. The resulting coherency stresses were determined by convergent beam electron diffraction (CBED) [19] and by analyzing the configuration of dislocations emitted from the lamellar interfaces [20]. The investigations have shown that the residual coherency stresses are comparable with the yield stress of the material. From theory it is expected that the residual coherency stresses present in the individual lamellae are inversely proportional to the lamellar spacing λ_L [21]. However, when sampled over a sufficiently large volume, the average of the coherency strain was zero. Thus, the sign of the coherency strain alternates from lamella to lamella. In the design of lamellar alloys λ_L is often reduced in order to maximize the yield stress. At the same time the coherency stresses grow both in absolute magnitude and relative to the yield stress [4]. Thus, in high-strength alloys the coherency stresses can be very large and can affect deformation, phase transformations, recovery, and recrystallization in various ways, as will be described in the following sections.

2.3 Modulated microstructures

In an attempt to improve the balance of mechanical properties, a novel type of TiAl alloys, designated γ-Md, with a composite-like microstructure has been recently developed [22]. The design bases on the general composition

$$\text{Ti-(40-44)Al-8.5 Nb.} \tag{4}$$

The characteristic constituents of the alloy are laths with a modulated substructure that is comprised of stable and metastable phases. The modulation occurs at the nanometer scale and thus provides an additional structural feature that refines the material. As indicated by X-ray analysis the constitution of the alloys (4) involves the β/B2, α_2 and γ phases. Additional X-ray reflections could be attributed to the presence of two orthorhombic phases with B19 structure, (oP4, Pmma and oC16, Cmcm). However, a clear association with the various orthorhombic structures reported in the literature was not possible because of their structural similarity. It might be expected that the evolution of the constitution does not reach thermodynamic equilibrium; thus, the number of the transformation products may be larger than expected from the phase rule. The microstructure of the alloy is shown in Fig. 5. The characteristic features are laths with a periodic variation in the diffraction contrast, which intersperse the other constituents. As shown in Fig. 5a, the contrast fluctuations occur at a very fine length scale. The evidence of the high-resolution electron microscope observations is that a single lath is subdivided into several regions with different crystalline structures with no sharp interface in between. The high-resolution micrograph in Fig. 5b shows a lath adjacent to a γ lamella in $<101]_\gamma$ projection, which can be used as a reference. The interface between the lath and the γ phase (designated as γ/T) consists mainly of flat terraces, which are parallel to the $(111)_\gamma$ plane. Steps of different heights often delineate the terraces. The modulated laths are comprised of an orthorhombic constituent, which is interspersed by slabs of β/B2 and a little α_2 phase (Fig. 6). Selected area diffraction of the orthorhombic constituent is consistent with the B19 phase, which can be described as the orthorhombic phase (oP4) or as a hexagonal superstructure of DO_{19} (hP8). In the Ti-Al system, the B19 structure has already been observed by Abe et al. [23] and Ducher et al. [24]. The B19 structure is structurally closely related to the orthorhombic phase (oC16, Cmcm) with the ideal stoichiometry Ti_2AlNb, which among the intermetallic compounds is remarkable for its relatively good room temperature ductility [25].

In Figs. 5b and 6a the B19 structure is imaged in the $[010]_{B19}$ projection. As deduced from the high-resolution images, the orientation relationships between the constituents involved in a modulated lath with the adjacent γ are [22]

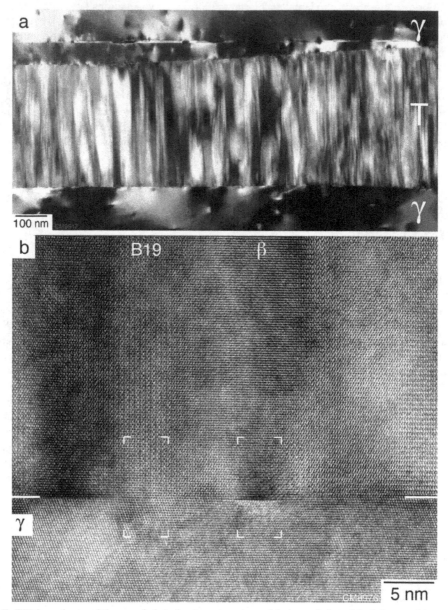

Fig. 5. TEM analysis of the modulated microstructure. (a) A modulated lath imaged by diffraction contrast. (b) High-resolution TEM micrograph of a modulated lath adjacent to a γ lamella.

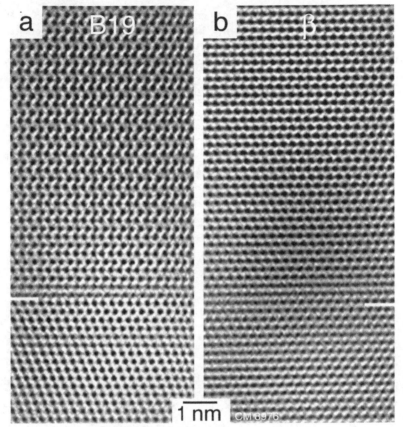

Fig. 6. High-resolution TEM evidence of the modulated microstructure. (a) and (b) Fourier-filtered images of the areas boxed in Fig. 5.

$$(100)_{B19} \mid\mid \{110\}_{\beta/B2} \mid\mid (0001)_{\alpha2} \mid\mid \{111\}_{\gamma}$$

$$[010]_{B19} \mid\mid <111>_{\beta/B2} \mid\mid <11\bar{2}0>_{\alpha2} \mid\mid <1\bar{1}0>_{\gamma}; [001]_{B19} \mid\mid <11\bar{2}>_{\gamma}. \tag{5}$$

In the diffraction pattern the presence of the periodic distortion is manifested by the existence of weak satellite reflections adjacent to the main reflections. The distance of the satellites from the main reflections is the reciprocal of the modulation wavelength, and the direction joining the satellites with their main reflections is parallel to the direction of the modulation vector. These observed features are reminiscent of a modulated structure, which in recent years have attracted considerable interest; for a review see [26]. A crystal structure is said to be modulated if it exhibits periodicities other than the Bravais lattice periodicities. These additional periodicities arise from one or more distortions, which increase to a maximum value and then decrease to the initial value. The modulation may involve atomic coordinates, occupancy factors or displacement parameters. The strain of the discontinuities is often relieved by a continuous and periodic variation of the physical properties of the product. First principle calculations of Nguyen-Manh and Pettifor [27] and Yoo and Fu [28] have shown that the β/B2 phase existing in TiAl alloys containing supersaturations of transition metals (Zr, V, Nb) is unstable under tetragonal distortion; a shear instability that was attributed to the anomalous (negative) tetragonal shear modulus. Specifically, B2 may transform by homogeneous shear to several low temperature orthorhombic phases, which can exist metastably. The energetically favourable transformations are [27]

$$B2(Pm3m) \rightarrow B19(Pmma) \rightarrow B33(Cmcm). \qquad (6)$$

At the atomic level, the B2 phase may transform to B19 by a shuffle displacement of neighboring $(011)_{B2}$ planes in opposite $[01\bar{1}]$ directions, as illustrated in Fig. 7. A subsequent displacement of neighboring $(011)_{B2}$ planes in the [100] direction generates the B33 structure [27]. In the system investigated here the predominant orthorhombic phase seems to be B19. It is tempting to speculate that the modulation of the laths is triggered by a periodic variation of the composition of the parent B2 phase, as occurs during spinodal decomposition. Clearly the mechanism requires further investigation.

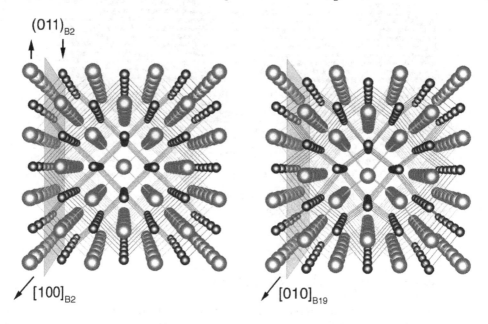

Fig. 7. Formation of the B19 structure from the parent B2 phase illustrated by perspective views of hard-sphere models. [100] projection of the B2 phase; arrow heads mark shuffle displacements of neighbouring $(011)_{B2}$ planes in opposite $[01\bar{1}]$ directions to form B19.

The modulated laths can apparently further transform into the γ phase, as demonstrated in Fig. 8. This process often starts at grain boundaries and proceeds through the formation of high ledges via distinct atomic shuffle displacements. As this transformation was frequently observed in deformed samples [22], it might be speculated that the process is stress induced and provides some kind of transformation toughening.

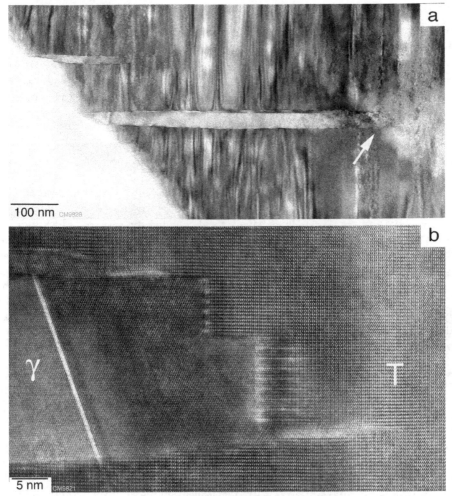

Fig. 8. Stress-induced transformation of a modulated lamella (T) into γ phase. (a) Generation of a γ lamella at a grain boundary. (b) Higher magnification of the area marked by the arrow in (a). Note the ledges at the interfaces.

3. Hot working

Titanium aluminide alloys are relatively brittle materials; attaining chemical homogeneity and refinement of the microstructure are therefore the most important prerequisites for engineering applications. To this end, large effort has been expended to establish wrought processing of TiAl alloys [29, 30]. Hot working of these alloys is generally impeded by a significant plastic anisotropy [31], low diffusivity [32] and the susceptibility to hot cracking [29]. While these aspects are well documented in the TiAl literature, there are many open questions about the elementary processes that determine dynamic recovery, recrystallization and phase transformations. These aspects are addressed in the present section.

3.1 The deformed state

A commonly held concept of hot working is that dynamic recovery and recrystallization are triggered by heterogeneities in the deformed state. Thus, the deformation mechanisms occurring in the majority phases γ(TiAl) and α_2(Ti$_3$Al) will briefly be discussed, for details see [1, 31]. Deformation of γ(TiAl) is mainly provided by ordinary dislocations with the Burgers vector b=1/2<110] and mechanical twinning along 1/6<$11\overline{2}$]{111}. Since twinning shear is unidirectional, the operating twinning systems vary with the sense of the load and the loading direction. There are crystal orientations for which twinning is forbidden. To a lesser extent deformation is provided by the motion of superdislocations with the Burgers vectors b=<101] and b=1/2<$11\overline{2}$]. The superdislocations exhibit an asymmetric non-planar core spreading; this results in a high glide resistance, which is sensitive to the direction of motion, for a review see [33]. Thus, the superdislocations do not move as readily as ordinary dislocations and mechanical twins. Taken together, this glide geometry gives rise to a significant plastic anisotropy of the γ phase. Ti$_3$Al alloys have several potential slip systems, which in principle may provide sufficient shear components for the deformation of polycrystalline material. However, there is a strong predominance for prismatic slip; plastic shear with c components of the hexagonal cell is practically impossible [34]. This plastic anisotropy appears to be even more enhanced at intermediate temperatures because the pyramidal slip systems exhibit an anomalous increase of the critical resolved shear stress with temperature. Thus, the brittleness of polycrystalline Ti$_3$Al alloys can be attributed to the lack of independent slip systems that can operate at comparable stresses; hence the von Mises criterion is not satisfied. Due to this situation deformation of γ(TiAl)+α_2(Ti$_3$Al) alloys is mainly carried by the γ phase.

There are various processes that might lead to heterogeneities in the deformed state. As a specific example for TiAl alloys, deformation heterogeneities resulting from mechanical twinning will be outlined in more detail. At the beginning of deformation the slip path of the twins is essentially identical with the domain size or lamellar spacing. As soon as multiple twinning with non-parallel shear vectors is activated, extensive intersections among twin bands occur. The intersection of a moving deformation twin with a barrier twin is expected to be difficult because the incorporation of an incident twinning system into the barrier twin may no longer constitute a crystallographically allowed twinning system. Several authors [35-39] have analyzed the mechanism and have proposed crystallographic relations by which an incident twin could intersect a barrier twin. The high-resolution electron micrograph shown in Fig. 9 demonstrates a so-called type-I twin intersection, which is favoured if the <$\overline{1}10$] intersection line is at 0° to 55° from the sample axis [35]. The structure is imaged along the common <$\overline{1}10$] direction of the two twins; this can be recognized by the different contrast of the (002) planes, which are alternately occupied by Ti and Al atoms. On its upper side the vertical twin is thicker than on its lower side, thus this twin was considered as incident twin T_i; T_b is the barrier twin. The intersection leads to a significant deflection of the two twins, which, however, is more pronounced for the barrier twin. The difficulty in forming a twin intersection is manifested by the ledged interface between the incident twin and the matrix (arrows 1 and 3) and various dislocations that emerge at the intersection zone (arrows 2). Figure 10 shows the details of the intersection zone. The intersection zone remains in the L1$_0$ structure and seems to be relatively free of defects; however highly defective regions border it. The (002)$_{Tb}$ planes of the central zone

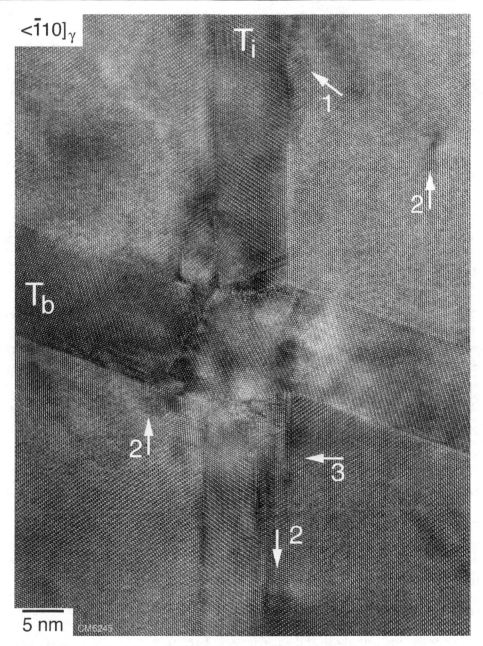

Fig. 9. Type-I intersection of two deformation twins observed after room temperature compression of a Ti-48.5Al-0.37C alloy. The structure is imaged along the common < $\bar{1}$10] intersection line. The vertical twin T_i is considered to be the incident twin because its upper side is thicker than its lower side; T_b is the barrier twin.

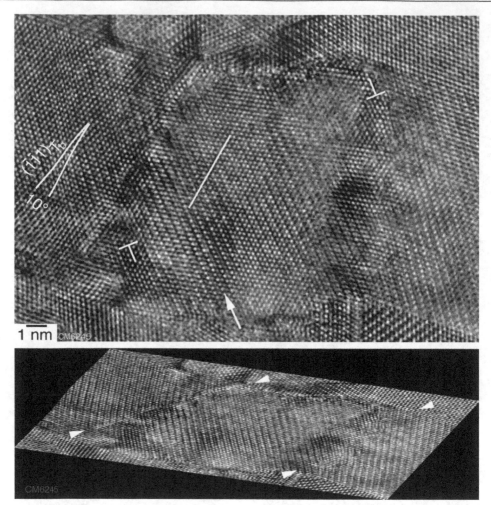

Fig. 10. Structural details of the intersection zone: translation of the twinning shear along the ($\overline{1}11$)$_\text{Tb}$ planes by 1/2<10$\overline{1}$] superpartials that become evident by extra (002)$_\text{Tb}$ planes. Two of the dislocations and their extra half planes are indicated by dislocation symbols. Note the anticlockwise rotation of the intersection zone with respect to the barrier twin by 10°, which is indicated by the traces of the respective ($\overline{1}11$)$_\text{Tb}$ planes. The compressed image below shows these features in more detail. Arrowheads mark the dislocation walls on either side of the intersection zone.

are not continuous with those of the barrier twin, but appear displaced along the ($\overline{1}11$)$_\text{Tb}$ planes of the barrier twin. This displacement is consistent with glide of 1/2<10$\overline{1}$] or 1/2<01$\overline{1}$] superpartials on ($\overline{1}11$)$_\text{Tb}$ planes. These dislocations become manifest by extra (002)$_\text{Tb}$ planes; a few of these dislocations and the orientation of their extra planes are indicated by dislocation symbols. It is tempting to speculate that these dislocations were generated under the high stress concentration acting at the corners of the intersection zone. It is worth adding that a shear accommodation by twinning along the $(\overline{1}11)_{Tb}$ planes is not possible because this would require anti-twinning operations. The close distance of these dislocations explains why the lattice of the intersection zone is not congruent with that of the barrier twin, but rotated by about 10° against the barrier twin. The image below is the micrograph compressed along the (002) planes and shows these details more clearly. The observation largely reflects the strong rotation field that was generated by the incident twin. The various dislocation reactions that could be involved in the intersection process probably give rise to dislocation emission; a few of these dislocations are marked with arrows 1 to 3. Twin intersections undoubtedly leave significant internal stresses and dense defect arrangements bordering the intersection zone. Under hot-working conditions these heterogeneities in the deformed state can be the prevalent sites for recrystallization. At elevated temperatures rearrangement of the dislocation walls surrounding the misoriented zone may occur by climb so that the misorientation with respect to the surrounding matrix increases. The intersection zone is transformed into a new grain of low internal energy, growing into deformed material from which it is finally separated by a high-angle boundary. This process is certainly driven by the release of stored energy. Thus, it might be expected that the structural heterogeneities produced by twin intersections act as precursor for recrystallization. Such processes are certainly beneficial for the conversion of the microstructure under hot-working conditions.

3.2 Dynamic recovery

There is good consensus that recovery and recrystallization are competing processes as both are driven by the stored energy of the deformed state. The extent of recovery is generally expected to depend on the stacking fault energy, which, in turn, determines dislocation dissociation. The ordinary dislocations, which mainly carry the deformation in γ(TiAl), have a compact core because dissociation would involve a high-energy complex stacking fault (CSF), which destroys the chemical environment of first neighbours in the fault plane. This compact core structure makes cross slip and climb of the ordinary dislocations relatively easy and is a good precondition for static and dynamic recovery. The most convincing evidence for climb of ordinary dislocations has also been obtained from in situ heating experiments performed inside the TEM [31]. The samples used were pre-deformed at room temperature to a strain of $\varepsilon=3\%$; this introduced sufficient dislocations for observation and certainly a small supersaturation of intrinsic point defects. Fig. 11 demonstrates the change of the dislocation fine structure occurring upon heating by a sequence of micrographs. The vacancies produced during room-temperature deformation condense onto screw dislocations, causing them to climb into helices. TEM observations performed after high-temperature deformation have also revealed a remarkable instability of twin structures [31]. Due to these factors, the release of strain energy by recovery is probably relatively easy and can account for the large reduction of the flow stress that has been observed after annealing at moderately high temperatures. Recovery lowers the driving force for recrystallization; thus, a significant amount of prior recovery may retard the recrystallization kinetics. More details about deformation and recovery are provided in [40].

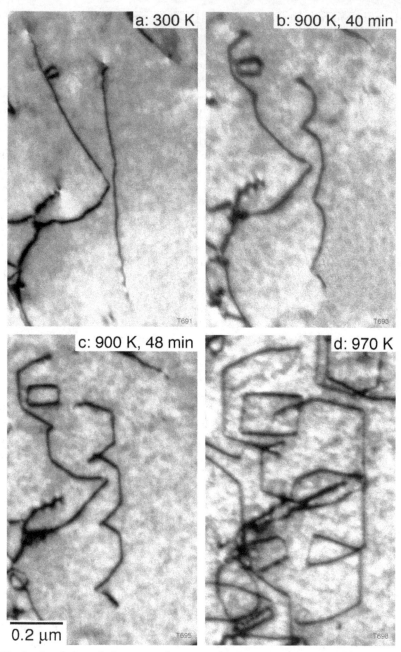

Fig. 11. Climb of ordinary dislocations during in situ heating inside the TEM at an acceleration voltage of 120 kV. Note the formation of a helical dislocation and the growth of prismatic dislocation loops. Cast Ti-48Al-2Cr, pre-deformation at room temperature in compression to ε=3%.

3.3 Dynamic recrystallization

The kinetics of dynamic recrystallization of TiAl alloys depends on several factors; these involve alloy composition, grain size of the starting microstructure, and hot working parameters [41]. The conversion of the coarse-grained lamellar ingot structure is probably the most difficult step in wrought processing. Thus, dynamic recrystallization occurring at this stage will be described as a specific example for TiAl alloys.

There is undisputed evidence that slip transfer through lamellar boundaries is difficult [42]. Plastic strain resulting from a cooperative operation of several deformation modes can be localized between the lamellar boundaries. These facts, combined with the flat plate geometry of the lamellae (Fig. 2), cause a marked plastic anisotropy of lamellar material, which affects the recrystallization behaviour [41]. The recrystallization kinetics is relatively fast if the lamellae orientation is parallel to the deformation axis. Deformation in this orientation apparently involves an element of instability, which is manifested by kinking of the lamellae (Fig. 12) and is reminiscent of buckling of load-carrying structures. In terms of a laminate model [43], the lamellar morphology may be considered as an ensemble of TiAl and Ti_3Al plates. When perfectly aligned with the compression axis, these "columns" are highly stable under compression, as long as the axial load is below a critical value. Above this critical load the equilibrium becomes unstable and the slightest disturbance will cause the structure to buckle. In the lamellar structure an upsetting moment might develop by lateral impinging of the lamellae by dislocation pile-ups or deformation twins. Furthermore, the elastic response of the α_2 and γ phase upon loading is significantly different. Thus, the tendency to instable buckling is expected to increase, if there is an inhomogeneous distribution of α_2 and γ lamellae. The process probably starts with local bending of the lamellae. From the curvature and thickness of the lamellae local strains can be deduced, which are often larger than 10 % and lead to the formation dense dislocation structures and of sub-boundaries (Figs. 13 and 14). Subsequent rotation and coalescence of these sub-grains occurs apparently in such a way that kinking of the lamellae is accomplished. Due to kinking, the lamellae are reoriented with respect to the deformation axis, which may support shearing along the lamellar interfaces. In the regions of highest local bending spheroidization and dissolution of the α_2 phase occur; this suggests that both the non-equilibrium constitution and the local stress provide the driving pressure for the observed phase transformation and recrystallization. All these aspects are manifested in Fig. 15, which shows the initial state of grain nucleation in a kinked α_2 lamella.

Micromechanical modelling [43, 44] has shown that polycrystalline lamellar material deforms very inhomogeneously. Localized stress concentrations have been recognized, which developed within the polycrystal upon straining due to the variations of grain size, shape and orientation. This led the authors to believe that constraints imposed by neighbouring grains are a dominant factor in determining the flow behaviour. Even under compression, very high tensile hydrostatic stresses are generated at the triple points of colony boundaries. There is a strong tendency to develop shear bands, kink bands, lattice rotations, and internal buckling of lamellae. The shear bands consist of extremely fine grains and may traverse the whole work piece, often resulting in gross failure. From a mechanical point of view, buckling failures do not depend on the yield strength of the material but only on the dimensions of the structure and the elastic properties of the material. In lamellar alloys of given lamellar spacing, the tendency to buckling and shear band formation increases with the axial length of plates. Thus, coarse-grained lamellar alloys seem to be prone to deformation instabilities and shear band formation, which turns out to be one of the prime problems in wrought processing of TiAl alloys.

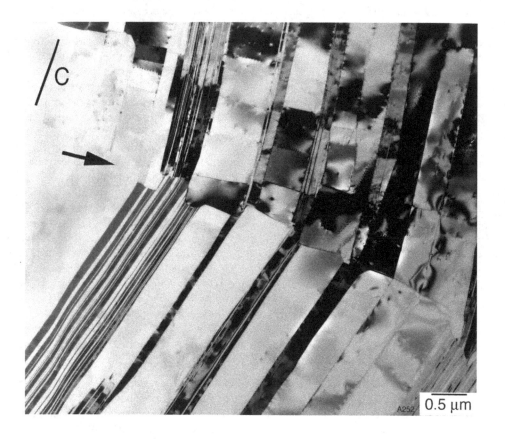

Fig. 12. Kinked lamellae in Ti-47Al-5Nb-0.2B-0.2C. The lamellar ingot material was subject at 1270 °C to a compression stress of $\sigma=(25.4\pm12.7)$ MPa fluctuating with 30 Hz. C indicates the orientation of the compression axis, which is parallel to the image plane.

Fig. 13. Sub-boundaries formed at kinked lamellae. Experimental details as for Fig. 12.

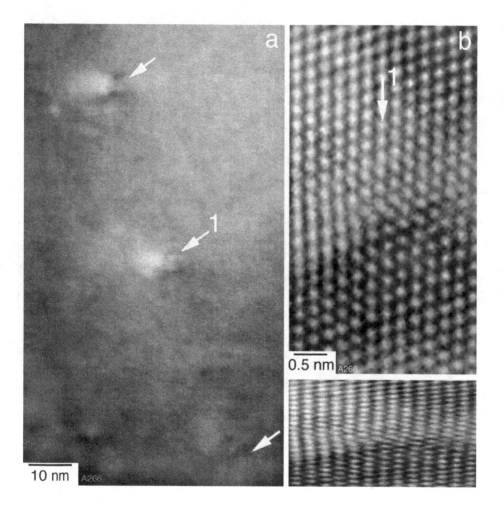

Fig. 14. Atomic structure of a sub-boundary formed at kinked γ lamella; experimental details as for Fig. 12. (a) Mixed ordinary dislocations situated in the sub-boundary (marked with arrow 1 in Fig. 13) and (b) on of the dislocations shown in higher magnification; the compressed image below shows the extra plane of the dislocation more clearly.

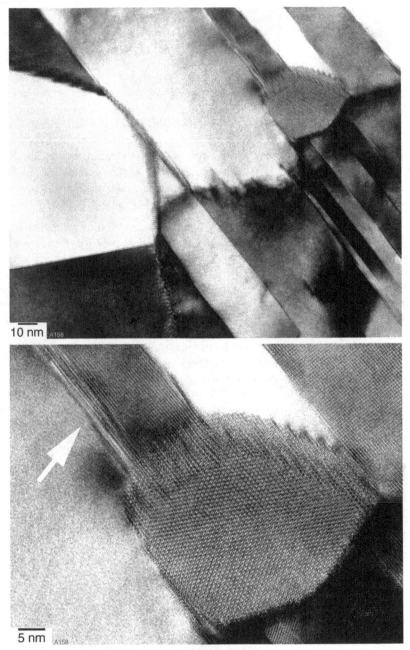

Fig. 15. Initial state of grain nucleation at kinked α_2 lamellae. The image below shows one of the grains in higher magnification. Note the shear processes occurring along the lamellae. Ti-45Al-8Nb-0.2C, sheet material; viewing direction in the transverse direction of the rolling plane.

4. Service-induced phase transformation and recrystallization

4.1 Creep

At temperatures above 650 °C damage of TiAl components may occur due to continuous creep. The relevant mechanisms are numerous and synergistic, depending on the operation conditions; for a review see [45]. Among the various microstructures that can be established in polycrystalline ($\alpha_2+\gamma$) alloys, fully lamellar alloys are most creep resistant. However, numerous investigations have demonstrated that the lamellar microstructure degrades upon creep. This structural instability is a serious problem for long-time service of lamellar alloys. In the following, a few examples of TEM analysis will be presented, which demonstrate the complexity of the processes involved in the degradation of the lamellar morphology. Most of these studies have been performed on samples that had been subject to long-term tensile creep at 700 °C under relatively low stresses of 80 to 140 MPa [46].

When compared with the microstructure of undeformed material, the lamellar interfaces in crept samples were highly imperfect. Figure 16a demonstrates the formation of a high interfacial step in a 60° pseudo twin boundary. The interfacial steps had often grown into broad zones, which extended over about 200 nm perpendicular to the interface; Fig. 16b demonstrates an intermediate stage of this growth process. Multiple-height ledges are commonly observed after phase transformation and growth, and several mechanisms have been proposed to explain the phenomenon [47]. Analogous to these models it is speculated that the large ledges observed in the crept TiAl alloy arise from one-plane steps, which moved under diffusional control along the interfaces and were piled up at misfit dislocations. A misfit dislocation with a Burgers vector component out of the interfacial plane is arrowed in Fig. 16a. Once a sharp pile up is formed, the configuration may rearrange into a tilt configuration with a long-range stress field. This would cause further perfect or Shockley partial dislocations to be incorporated into the ledge and would also explain that in all cases the ledges were associated with misfit dislocations. The detailed atomic structure of the macro ledges is not clear. As can be seen in Figs. 16, there is a variation of the contrast in the ledges with a periodicity of three (111) planes. This is reminiscent of the 9R structure, which is a phase that probably has a slightly higher energy than the $L1_0$ ground state. The formation of the 9R structure is a well-known phenomenon in many f.c.c. metals that exhibit twinning. Singh and Howe [48] have recognized the 9R structure in heavily deformed TiAl. It should be noted, however, that similar three-plane structures have been observed in a massively transformed, but undeformed Ti-48.7Al [49]. In this work the contrast phenomena have been interpreted as arising from overlapping twin related γ variants. Nevertheless, the macro-ledges are a characteristic feature in the microstructure of crept samples and represent at least a highly faulted $L1_0$ structure; the question is only whether there is a periodicity in the fault arrangement. When the macro-ledges grow further, it might be energetically favourable to reconstruct the $L1_0$ structure and to nucleate a new γ grain. Figure 17 probably demonstrates an early stage of such a process. The recrystallized grains usually have a certain orientation relationship with respect to the parent γ lamellae; Fig. 18 indicates that the (001) planes of the recrystallized grain are parallel to the ($1\overline{1}1$) planes in the parent lamella γ_1. There is a significant mismatch for this orientation relationship, which is manifested by a high density of ledges and dislocations at the $(001)||(1\overline{1}1)$ interface. Recrystallization of ordered structures has been investigated in several studies [50, 51]. There is a drastic reduction in grain boundary mobility, when compared with disordered metals. Recovery of ordered alloys is also complicated by the fact that the ordered state has to be restored. In this respect it is interesting to note that the small grain shown in Fig. 18 is completely ordered giving the impression that the ordering is immediately established after grain nucleation or that nucleation occurred in the ordered state. This might be a consequence of the fine scale of the lamellar microstructure and the heterogeneous grain nucleation at the interfacial ledges. There are certainly crystallographic constraints exerted by the parent lamellae adjacent to the ledges, which may control nucleation and growth. Clearly, the process needs further investigation.

There is a significant body of evidence in the TiAl literature indicating that dissolution of α_2 lamellae occurs during creep [45, 46]. The phase transformation is probably driven by a non-equilibrium constitution. High-temperature creep is expected to promote phase transformation towards equilibrium constitution; thus, dissolution of α_2 and formation of γ occurs [52]. The high-resolution micrograph shown in Fig. 19 supports this reasoning; there is clear evidence that the density of steps at the α_2/γ interfaces is significantly higher than that at the γ/γ interfaces, meaning that the α_2 lamella dissolves, whereas the γ lamellae are relatively stable. The processes eventually end with the formation of new grains (Fig. 20) and a more or less complete conversion of the lamellar morphology into a fine spheroidized microstructure. The $\alpha_2 \rightarrow \gamma$ phase transformation is often associated with local deformation, as suggested by Fig. 21. The micrograph shows two α_2 terminations that are connected by an interface, thus the α_2 lamella is partially dissolved. Two twins were emitted at one of the terminations and a dislocation with a Burgers vector out of the interface is present (Fig. 22). The α_2 terminations have extremely small principal radii of curvature. The elimination of such structural features reduces the surface energy and provides a driving force towards further coarsening. The interface connecting the two α_2 terminations exhibits a fault translation that corresponds to an intrinsic stacking fault. This is indicated by the stacking sequence $ABC\ B\ CAB$. The nature of the interface between two phases is determined by their structural relationships. There is a strong tendency for planes and directions with the highest atomic densities to align across the interface. As suggested by Chalmers and Gleiter [53], a better atomic fit at a boundary could result if atoms were moved away from coincident sites by a rigid-body displacement of one grain relative to the other by a constant displacement vector. Atomic modelling performed in this context (for a review see [33]) has shown that the energy of lamellar interfaces could be minimized by a rigid-body translation along the vectors $f_{APB}=1/2<10\bar{1}]_\gamma$, $f_{SISF}=1/6[11\bar{2}]_\gamma$ and $f_{CSF}=1/6[\bar{2}11]_\gamma$. These translations correspond to the formation of an antiphase boundary (APB), a superlattice intrinsic stacking fault (SISF) and a complex stacking fault (CSF), respectively, at the interface. In the present case the translation vector is of type $f_{SISF}=1/6[11\bar{2}]_\gamma$. The observation underlines once again the fine scale of interface processes that may occur during creep of lamellar alloys.

The $\alpha_2 \rightarrow \gamma$ transformation requires a change of both the stacking sequence and the local composition. However, achieving the appropriate composition requires long-range diffusion, which at a creep temperature of 700 °C is very sluggish. Low-temperature diffusion might be supported by the presence of Ti_{Al} antisite defects [6, 32]. In the newly formed γ phase a high density of such defects is certainly formed, in order to accommodate the excess of titanium. Thus, a substantial antistructural disorder occurs, which forms a percolating substructure. Under such conditions, the antisite defects may significantly contribute to diffusion because antistructural bridges (ASB's) are formed. An elementary bridge event involving one vacancy and one antisite defect consists of two nearest neighbour jumps, which result in a nearest neighbour displacement of two atoms of the same species. For the $\alpha_2 \rightarrow \gamma$ transformation the so-called ASB-2 mechanism [32] might be relevant, which requires only low migration energy. Diffusion may also be supported by the mismatch structures present at the interfaces. Dislocations and ledges represent regions where the deviation from the ideal crystalline structure is concentrated. These are paths of easy diffusion, which can effectively support the exchange of Ti and Al atoms. One may expect all these processes to be thermally activated und supported by superimposed external stresses. In this respect, the coherency stresses present at the interfaces are certainly significant because they are comparable to or even higher than the shear stresses applied during creep tests and are often associated with mismatch structures. Thus, given a non-equilibrium constitution, it is understandable that drop in the α_2 volume content also occurs, when a lamellar alloy is subject to the same temperature/time profile without externally applied stress [54].

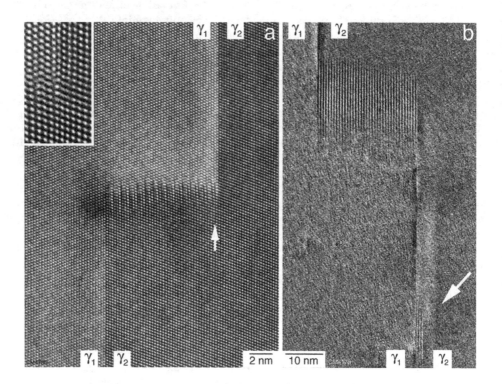

Fig. 16. Degradation of the lamellar structure in Ti-48Al-2Cr under long-term creep at T=700 °C, σ_a=140 MPa, for t=5988 hours to strain ε=0.69 %. (a) Formation of an interfacial step in a 60° pseudo twin boundary. Note the interfacial dislocation (arrowed) that is manifested by an additional {111} plane. (b) A macro-ledge present in a 60° pseudo twin boundary.

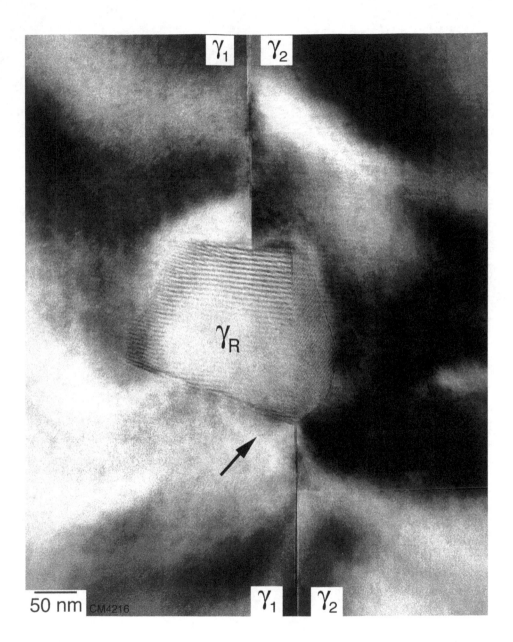

Fig. 17. Recrystallized grain γ_R formed at a ledge in a lamellar interface joining the gamma variants γ_1 and γ_2 with a pseudo twin orientation relationship. Note the step in the interface and the ordered state of the recrystallized grain. Experimental details as for Fig. 16.

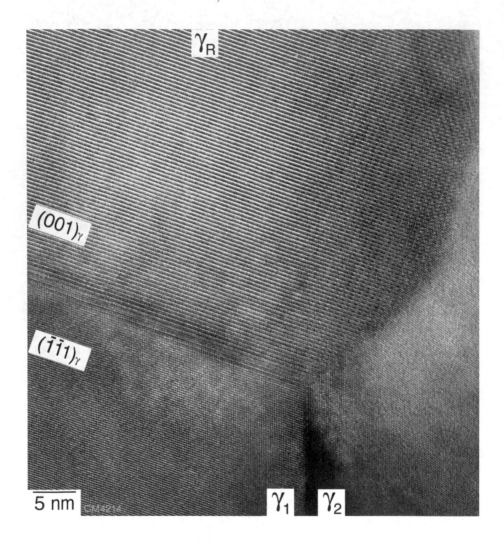

Fig. 18. Higher magnification of the boundary triple-point marked in Fig. 17. Note the orientation relationship (001) | | (111) between the recrystallized grain γ_R and lamella γ_1.

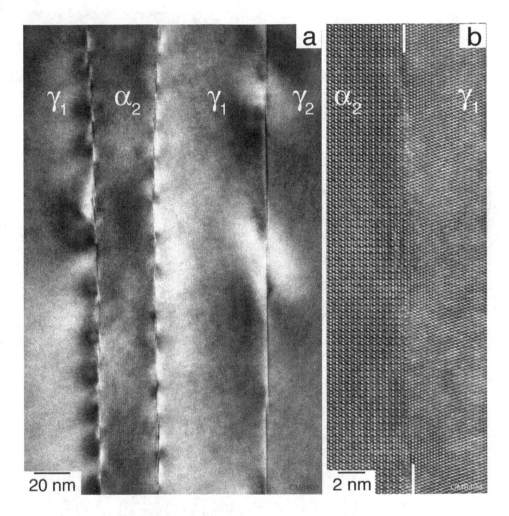

Fig. 19. Initial stage of the $\alpha_2 \rightarrow \gamma$ phase transformation in the lamellar structure of a Ti-48Al-2Cr alloy occurring during long-term creep at T=700 °C, σ_a=110 MPa, t=13400 hours to strain ε=0.46 %. (a) Low-magnification high-resolution image of the lamellar structure. Note the significantly higher density of steps at the α_2/γ interfaces, which indicates dissolution of α_2 phase. (b) Atomic structure of one of the α_2/γ interfaces demonstrating its stepped character.

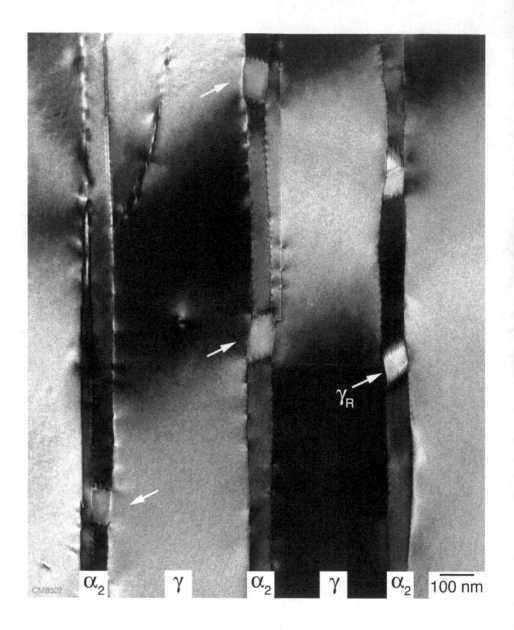

Fig. 20. Spheroidization of α_2 lamellae due to the formation of γ grains (arrowed, designated as γ_R). Experimental details as for Fig. 19.

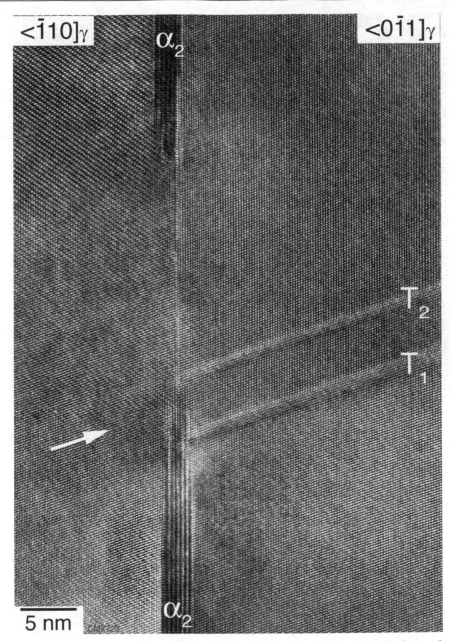

Fig. 21. A partially dissolved α_2 lamella in Ti-46.5Al-4(Cr, Nb, Ta, B) embedded in γ phase. Creep deformation at T=700 °C, σ_a=200 MPa to strain ε=1.35 %. Note the two α_2 terminations that are connected by an interface and the emissions of two twins T_1 and T_2 at one of the terminations. The stacking sequence indicates a rigid body translation of the adjacent γ lamellae.

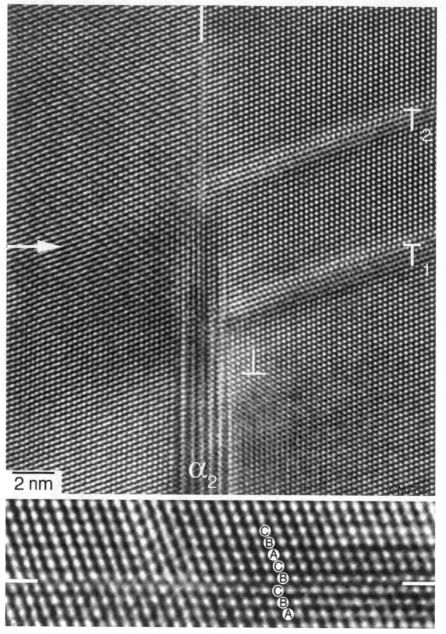

Fig. 22. Higher magnification of the area marked in Fig. 21 showing deformation activity in the vicinity one of the α_2 terminations. Two mechanical twins and a dislocation with a Burgers vector out of the interface plane (marked by symbol) are present. The lower image shows the stacking sequence across the interface (parallel to the arrow) that connects the two α_2 terminations. The stacking sequence indicates the presence of a fault translation.

4.2 Fatigue

By far, the most anticipated engineering applications of TiAl alloys involve components that are subjected to fluctuating or cyclic loading. The capability of TiAl alloys to sustain such loading conditions is inherently limited by the plastic anisotropy at the dislocation level and the lack of independent slip and twinning systems that can operate under reversed plastic straining. Low-cycle fatigue (LCF) is a progressive failure phenomenon brought about by cyclic strains that extend into the plastic range. Thus, the LCF life is largely determined by the amount of inelastic strain in each cycle. While the macroscopic LCF phenomena are well characterized [55], it is only recently that information about the structural degradation occurring upon fatigue has been obtained [56]. A few examples for these processes will be demonstrated in this section.

The fatigue study was performed on an extruded Ti-45Al-8Nb-0.2C alloy (TNB-V2), which contains a significant amount of β/B2 phase and an orthorhombic phase with B19 structure. TEM examination performed after room temperature fatigue has shown that the B19 structure transforms into γ phase. The phase transformation occurs in such a way that extremely fine shear bands are formed (Fig. 23). The result is a lamellar morphology that is comprised of extremely fine γ lamellae adjacent to B19 phase (Fig. 23b). A general observation is that two different γ variants are usually generated adjacent to the B19 phase, as seen for almost all the γ lamellae in Fig. 23b. If, for example, the stacking sequence of the $\{111\}_\gamma$ planes of the variant γ_1 is labelled *ABC*, that of variant γ_2 is *CBA*. This inversion of the stacking sequence is thought to induce strain fields of opposite signs, which eventually reduces the total strain energy. It might be speculated that such a combination of shear processes makes the transformation easier. The transformation often starts at grain boundaries and proceeds through the formation of ledges via distinct atomic shuffle displacements; these details are shown in Fig. 24. It should be noted that the B19→γ transformation has been observed at all the fatigue test temperatures investigated. Due to the difference in lattice constants between the γ and B19 phases the transformation may accommodate local strains and is thus expected to serve as a toughening mechanism. Nevertheless, the life of TiAl testpieces under low cycle fatigue with plastic cyclic strain amplitudes of a few tenth of a percent is limited to a several hundred cycles.

5. Diffusion bonding

Solid-state diffusion bonding provides a means of joining TiAl alloys without melting of the base materials. The principal diffusion bonding parameters, temperature and stress, depend on yield strength, work hardening behaviour and creep resistance of the mating alloy coupons. The bonding conditions should be chosen so that coalescence of contacting surfaces is produced by asperity deformation, but without gross deformation of the component. The effects of bonding temperature and bonding stress are synergistic; at higher temperature less stress is required and vice versa. For more details see [57].

During diffusion bonding of (α_2+γ) alloys a three layer process zone is typically developed that involves a fine grained layer of α_2 phase at the former contact plane of the diffusion couple, a region made up of relatively large recrystallized grains, followed by a region of deformed bulk material (Fig. 25). The bond layer consists of fine stress free grains (Fig. 26), which were identified by EDX and EBSD analysis as α_2 phase. It is well documented in the

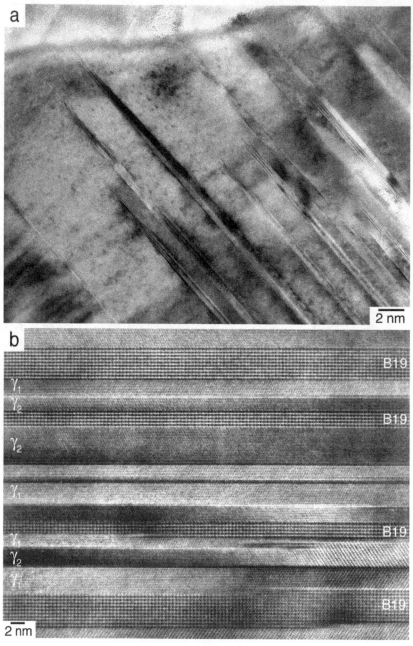

Fig. 23. Phase transformation B19→γ during low cycle fatigue at T=550 °C, $\Delta\varepsilon_t/2=\pm0.7$ %, N_f=452; nearly lamellar Ti-45Al-8Nb-0.2C. (a) Low-magnification high-resolution TEM micrograph showing fine γ lamellae produced in the B19 phase. (b) Lamellar morphology consisting of fine γ lamellae adjacent to B19 phase.

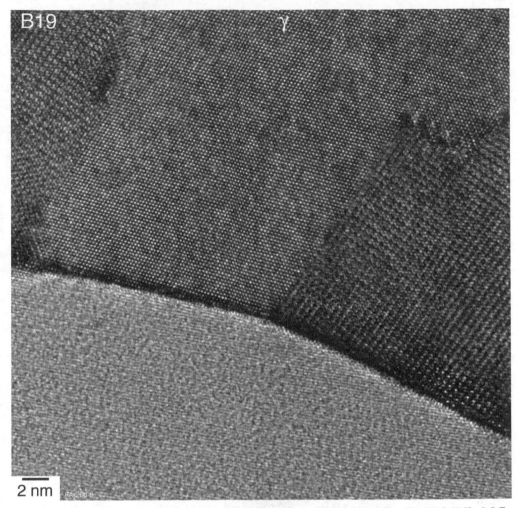

Fig. 24. Transformation B19→γ near to a grain boundary. Nearly lamellar Ti-45Al-8Nb-0.2C, sample fatigued at 25 °C to failure after N=641 cycles with R=-1 and a total strain amplitude $\Delta\varepsilon_t/2$=0.7 %.

literature [58] that a very small amount of oxygen can stabilize the α_2 phase. This finding is consistent with an early investigation of Godfrey et al. [59] performed on diffusion bonded Ti-48Al-2Mn-2Nb. Orientation analysis performed on the α_2 grains has shown that most of the newly formed α_2 grains have an orientation that is suitable for prismatic glide. This data could reflect the well-known plastic anisotropy of the α_2 phase, according to which the activation of prismatic slip along $1/3<11\overline{2}0>\{10\overline{1}0\}$ is by far easiest (Sect. 3.1). It might be speculated that nucleation and growth of new α_2 grains are controlled by the deformation constraints operating during bonding in that the preferred grain orientation ensures strain accommodation on the most favourably slip system.

The α_2 phase at the bond layer is formed at the expense of the Ti content of the adjacent regions, which needs long-range diffusion. The Ti transport could be supported by the anti-structural disorder, as already mentioned in Sect. 4.1. Furthermore, Ti transport could be accomplished by diffusion along the various internal boundaries present in the fine-grained materials. In this respect the lamellar interfaces are probably important because the deviation from the ideal crystal structure occurs and dense arrangements of misfit dislocations are present. Due to the transport processes described, the Ti content of the pre-existing α_2 and β phases situated next to the bonding layer gradually decreases and eventually falls below the critical composition required for their existence; eventually, these phases transform into $\gamma(TiAl)$. The effect is most pronounced in lamellar colonies that are in contact with the bonding line, presumably because pipe diffusion along the lamellar interfaces is significant. A common observation supporting this mechanism is that the newly formed α_2 grains are connected with pre-existing α_2 lamellae, which in a sense feed the chemically driven generation of α_2 phase at the bonding layer. All in all, the process zone of diffusion bonded TiAl couples reflects the combined influences of chemical driving pressure due to the oxygen contamination at the bonding surfaces and dynamic recrystallization induced by asperity surface deformation.

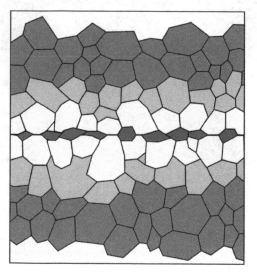

former contact zone (BL)
dynamic recrystallization (DRX)
remnant deformation (RD)

bulk structure

Fig. 25. Schematic illustration of the process zone observed after diffusion bonding at relatively low stresses. BL - fine grained bond layer at the former contact plane of the diffusion couple, DRX - region consisting of relatively large recrystallized grains, RD - initial bulk material but with remnant plastic deformation.

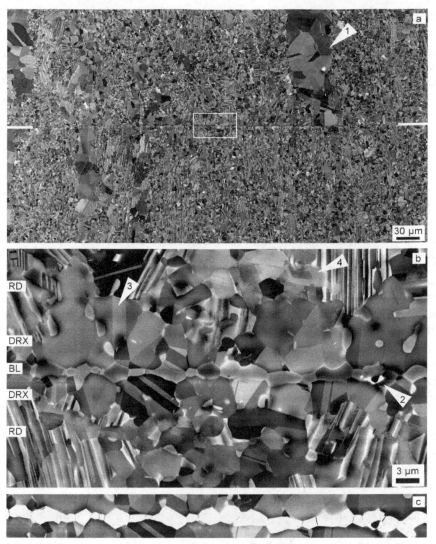

Fig. 26. Cross section of a bond in Ti-46.5Al formed at T= 1273K, σ=20 MPa and t=2h. Scanning electron micrographs taken in the backscattered mode. (a) Low magnification image showing the gross structure of the bond; the horizontally oriented bonding layer is marked by white bars. Note the inhomogeneity of the starting material that is manifested by a banded structure parallel to the extrusion direction (vertical in the micrograph) involving remnant lamellae, fine-grained regions and large γ grains (arrow 1). (b) Higher magnification of the area boxed in (a) showing the bonding layer BL, the recrystallized region DRX and the region with remnant deformation RD in more detail. Note the pore in the bonding layer (arrow 2) and the annealing twins in the γ grains of the DRX region (arrows 3 and 4). (c) Bonding layer consisting of fine-grained α2 phase as identified by EBSD and EDX analysis [57].

6. Shot peening

Shot peening is a cold working process in which the surface of a component is blasted with small spherical media called shot. A compressive layer is formed by a combination of subsurface compression developed at the Hertzian impression combined with lateral displacement of the surface material around each of the dimples formed. Since fatigue cracks will not initiate nor propagate in compressively stressed regions, shot peening can greatly enhance fatigue life.

In TiAl alloys shot peening produces a heavily deformed surface layer with a thickness of 10 μm to 80 μm depending on the microstructure and yield stress of the substrate alloy [60]. Deformation is characterized by intensive glide and mechanical twinning, involving all potential slip systems available in the major phases $\alpha_2(Ti_3Al)$ and $\gamma(TiAl)$, not only the easy ones. On the mesoscopic scale, buckling and kinking of the lamellae manifest deformation. TEM observations have revealed a remarkable conversion of the microstructure involving dynamic recrystallization and $\alpha_2 \rightarrow \gamma$ phase transformation. The phase transformation starts with a splitting of α_2 lamellae at positions of strong bending or kinking, where the elastic stresses are highest. Another prominent damage mechanism is amorphisation. As shown in Fig. 27a, nano-crystalline grains are embedded in an almost featureless amorphous phase; Fig. 27b demonstrates the gradual loss of crystallinity towards to the adjacent amorphous phase. There seems to be a significant mismatch between the crystalline and amorphous phases, which is indicated by a systematic array of like dislocations (Fig. 28). From the dislocation separation distance it may be concluded that the mismatch is at least 2 % to 5 %.

The observation of an amorphous phase is surprising. However, there are a few arguments that make its existence plausible. Firstly, in the surface layer the material undergoes severe plastic deformation. This introduces various defects, raises the free energy, and creates fresh surfaces due to the formation of slip steps and localized cracking. Secondly, there is certainly a substantial pick up of nitrogen, oxygen, and perhaps hydrogen because the shot peening was performed in air. Thus, several metastable nitride, oxide and hydride phases can be formed. The presence of these interstitial elements in the α_2 and γ phases may favour their amorphisation. Unfortunately, the nature of the crystalline grains embedded into the amorphous phase could not be determined, but it might be speculated that they are oxide, nitride or hydride phases. The formation of one of these crystalline surface phases could be an intermediate state before amorphisation eventually starts. This is suggested by two observations. The crystalline surface phase contains a high density of dislocations, which are often arranged in dipole or multipole configurations. The compressed image reveals significant bending of the lattice planes, which suggests high internal stresses. Another remarkable feature of the crystalline surface phase is antiphase boundaries (APB), as seen in Fig. 29. The APB's provide a local loss of order and are often associated with adjacent amorphous phase. Thus, it is speculated that the formation of APB's represents the initial stage of amorphisation.

Taken together, theses factors apparently make the crystalline surface phase prone to further structural changes, which could be directly observed in the electron microscope. Figure 30 demonstrates the transformation of the crystalline phase by a couple of micrographs. The slurry contrast in the micrograph on the left hand side indicates that the structure was about to transform into the structure shown on the right hand side. The observed conversion of the structure usually involved a very small volume of several ten nanometres and often occurred within a few ten seconds.

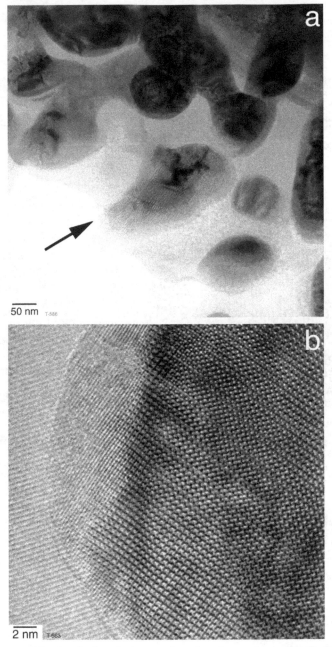

Fig. 27. Partial amorphisation in the shot peened surface layer. Lamellar Ti-45Al-10Nb, shot peened at room temperature with an Almen intensity of 0.4 mm N. (a) Crystalline grains embedded in amorphous phase. (b) Gradual loss of crystallinity of a grain adjacent to the amorphous phase.

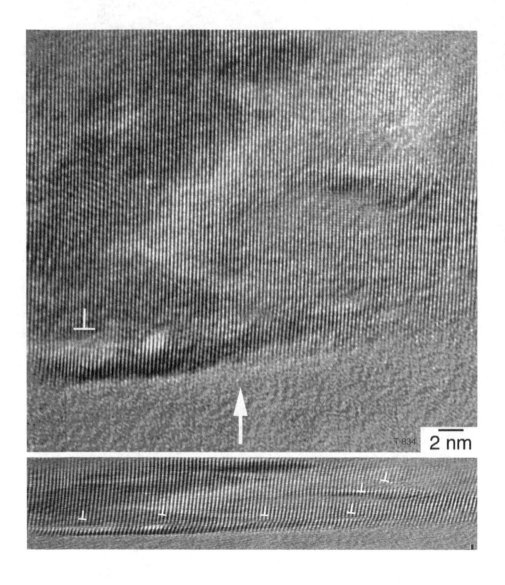

Fig. 28. Misfit between the amorphous and crystalline surface phases produced by shot peening. Note the arrangement of like dislocations at the interface. The compressed image below shows these details more clearly. Experimental conditions as for Fig. 27.

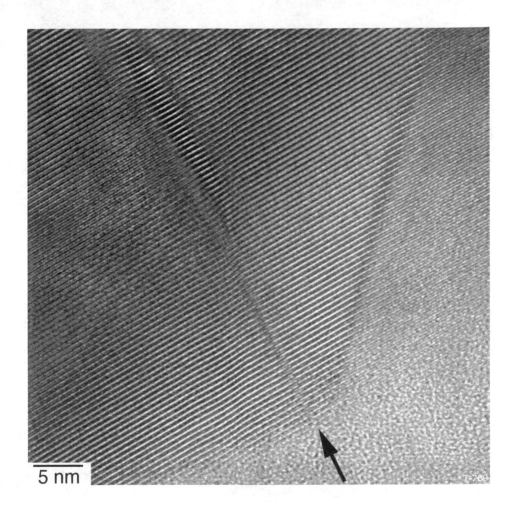

Fig. 29. An antiphase boundary (arrowed) in a crystalline grain embedded into the amorphous surface layer produced by shot peening. Experimental conditions as for Fig. 27.

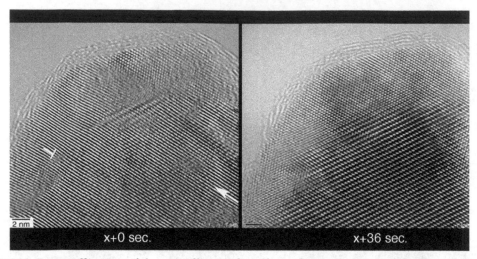

Fig. 30. Recrystallization of the crystalline surface phase observed in situ in a high-resolution transmission microscope (acceleration voltage 300 kV). Experimental conditions as for Fig. 27.

7. Conclusions

As with other metals recrystallization in multiphase titanium aluminide alloys is triggered by heterogeneities in the deformed state. However, there are several specifics that involve the following features.

The deformation heterogeneities are formed by a significant plastic anisotropy of the majority phases γ(TiAl) and α_2(Ti$_3$Al), twin intersections and elastic buckling of lamellar constituents.

The release of strain energy by recovery seems to be relatively easy and probably retards the recrystallization kinetics.

In the evolution of the microstructure concurrent phase transformations occur that are primarily initiated by a non-equilibrium constitution; other factors supporting phase transformations are local mechanical stresses or chemical driving pressure due contamination with gaseous elements. The combination of these effects can give rise to solid state amorphisation.

The complex interactions between deformation, recovery and phase transformation are of great importance for processing and service of TiAl alloys.

8. Acknowledgments

The author acknowledges the continuous support of St. Eggert, D. Herrmann, U. Lorenz, M. Oehring, and J. Paul, from the Helmholtz Zentrum Geesthacht, Germany. Thanks are due to Th. Heckel, A. El-Chaikh and H.-J. Christ from Universität Siegen, Germany, for performing the fatigue experiments. The financial support by the Deutsche Forschungsgemeinschaft (Projects AP 49/5 and AP 49/4-6) is gratefully acknowledged.

8. References

[1] F. Appel, J.D.H. Paul and Michael Oehring, *Gamma Titanium Aluminide Alloys - Science and Technology* (Wiley VCH, Weinheim, 2011).

[2] M. Yamaguchi, H. Inui and K. Ito, Acta Mater. 47, 307 (2000).

[3] C. McCullough, J.J. Valencia, C.G. Levi, and R. Mehrabian, *Acta Metall.* 37, 1321 (1989).

[4] P. M. Hazzledine, Intermetallics 6, 673 (1998).

[5] M.J. Blackburn, in: R.I. Jaffe, N.E. Promisel, eds. *The Science Technology and Applications of Titanium*, Pergamon, Oxford 1970.

[6] U. Fröbel and F. Appel, Acta Mater. 50, 3693 (2002).

[7] H.I. Aaronson, Metall. Trans. A, 24A, 241 (1993).

[8] F.C. Frank and J.H. van der Merwe, Proc. Roy. Soc. Series A - Mathematical and Physical Sciences A 198, 205 (1949).

[9] M.G. Hall, H.I. Aaronson and K.R. Kinsman, Surf. Sci. 31, 257 (1972).

[10] R.C. Pond, in: *Dislocations in Solids, Vol. 8*, ed. F.R.N. Nabarro (North-Holland, Amsterdam, 1989), p.1.

[11] J.M. Howe, R.C. Pond and J.P. Hirth, Progr. Mater. Sci. 54, 792 (2009).

[12] L. Zhao and K. Tangri, Acta Metall. Mater. 39, 2209 (1991)

[13] S.R. Singh and J. M. Howe, Phil. Mag. A 66, 739 (1992).

[14] S. Rao, C. Woodward and P. Hazzledine, in: *Defect Interface Interactions*, Materials Research Society Symposium Proceedings, Vol. 319, eds. E.P. Kvam, A.H. King, M.J. Mills, T.D. Sands, and V. Vitek (MRS, Pittsburgh, PA, 1994), p. 285.

[15] P. Shang, T.T. Cheng and M. Aindow, Phil. Mag. A 79, 2553 (1999).

[16] R.C. Pond, P. Shang, T.T. Cheng, and M. Aindow, Acta Mater. 48, 1047 (2000).

[17] P. Shang, T.T. Cheng and M. Aindow, Phil. Mag. Lett. 80, 1 (2000).

[18] J.P. Hirth and J. Lothe, *Theory of Dislocations* (Krieger, Melbourne, 1992).

[19] P.M. Hazzledine, B.K. Kad, H.L. Fraser, and D.M. Dimiduk, in: *Intermetallic Matrix Composites II*, Mater. Res. Soc. Symp. Proc. Vol. 273, eds. D.B. Miracle, D.L. Anton, J.A. Graves (MRS, Pittsburgh, PA, 1992), p.81.

[20] F. Appel and U. Christoph, Intermetallics 7, 1173 (1999).

[21] B. Shoykhet, M.A. Grinfeld and P.M. Hazzledine, Acta Mater. 46, 3761 (1998).

[22] F. Appel, J.D.H. Paul and M. Oehring, Mater. Sci. Eng. A 493, 232 (2008).

[23] E. Abe, T. Kumagai and M. Nakamura, Intermetallics 4, 327 (1996).

[24] R. Ducher, B. Viguier and J. Lacaze, Scripta Mater. 47, 307 (2002).

[25] A.K. Gogia, T.K. Nandy, D. Banerjee, T. Carisey, J.L. Strudel and J.M. Franchet, Intermetallics 6, 741 (1998).

[26] T. Haibach and W. Steurer, Acta Crystallographica A, 52, 277 (1996).

[27] D. Nguyen-Manh and D.G. Pettifor, in: *Gamma Titanium Aluminides 1999*, eds. Y-W. Kim, D.M. Dimiduk and M.H. Loretto (TMS, Warrendale, PA, 1999), p. 175.

[28] M.H. Yoo and J. Zou, C.L. Fu, Mater. Sci. Eng. A 192-193, 14 (1995).

[29] V. Seetharaman and S.L. Semiatin, Metall. Trans. A, 27A, 1987 (1996).

[30] Y-W. Kim and D.M. Dimiduk, in: *Structural Intermetallics 1997*, eds. M.V. Nathal, R. Darolia, C.T. Liu, P.L. Martin, D.B. Miracle, R. Wagner, and M. Yamaguchi (TMS, Warrendale, PA, 1997), p. 531.

[31] F. Appel and R. Wagner, Mater. Sci. Eng. R22, 187 (1998).

[32] Y. Mishin and Chr. Herzig, Acta Mater. 48, 589 (2000).

[33] M.H. Yoo and C.L. Fu, Metall. Mater. Trans. A, 29A, 49 (1998).

[34] Y. Minonishi, Mater. Sci. Eng. A 192-193, 830 (1995).

[35] Y.Q. Sun, P.M. Hazzledine and J.W. Christian, Phil. Mag. A 68, 471 (1993).

[36] Y.Q. Sun, P.M. Hazzledine and J.W. Christian, Phil. Mag. A 68, 495 (1993).

[37] S. Wardle, I. Phan and G. Hug, Phil. Mag. A 67, 497 (1993).

[38] J.W. Christian and S. Mahajan, Progr. Mater. Sci. 39, 1 (1995).

[39] F. Appel, Phil. Mag. 85, 205 (2005).

[40] F. Appel, U. Sparka and R. Wagner, Intermetallics 7, 325 (1999).

[41] R. M. Imayev, V.M. Imayev, M. Oehring, and F. Appel, Metall. Mater. Trans. A, 36A, 859 (2005).

[42] T. Fujiwara, A. Nakamura, M. Hosomi, S.R. Nishitani, Y. Shirai, and M. Yamaguchi, Phil. Mag. A, 61, 591 (1990).

[43] Th. Schaden, F.D.Fischer, H. Clemens, F. Appel, and A. Bartels, Adv. Eng. Mater. 8, 1109 (2006).

[44] R.A. Brockman, Int. J. Plasticity 19, 1749 (2003).

[45] J. Beddoes, W. Wallace and L. Zhao, Int. Mater. Rev. 40, 197 (1995).

[46] M. Oehring, F. Appel, P.J. Ennis, and R. Wagner, Intermetallics 7, 335 (1999).

[47] T. Furuhara, J.M. Howe and H.J. Anderson, Acta Metall. Mater. 39, 2873 (1991).

[48] S.R. Singh and J.M. Howe, Phil. Mag. Lett. 65, 233 (1992).

[49] E. Abe, S. Kajiwara, T. Kumagai, and N. Nakamura, Phil. Mag. A 75, 975 (1997).

[50] R.W. Cahn, in: *High Temperature Aluminides and Intermetallics*, eds. S.H. Whang, C.T. Liu, D.P. Pope, and J.O. Stiegler (TMS, Warrendale, PA, 1990), p. 245.

[51] F.J. Humphreys and M. Hatherly, *Recrystallization and Related Annealing Phenomena* (Pergamon, Oxford, 1995).

[52] F. Appel, U. Christoph and M. Oehring, Mater. Sci. Eng. A 329-331, 780 (2002).

[53] B. Chalmers and H. Gleiter, Phil. Mag. 23, 1541 (1971).

[54] D. Hu, A.B. Godfrey and M. Loretto, Intermetallics 6, 413 (1998).

[55] G. Hénaff and A.-L. Gloanec, Intermetallics 13, 543 (2005).

[56] F. Appel, Th. Heckel and H.J. Christ, Int. J. Fatigue 32, 792 (2010).

[57] D. Herrmann and F. Appel, Metall. Mater. Trans. A, 40A, 1881 (2009).

[58] U.R. Kattner, J.C. Liu and Y.A. Chang, Metall. Trans. A, 23A, 2081 (1992).

[59] S.P. Godfrey, P.L. Threadgill and M. Strangwood, in: *High-Temperature Ordered Intermetallic Alloys VI*, Materials Research Society Symposia Proceedings, Vol. 364, eds. J.A. Horton, I. Baker, S. Hanada, R.D. Noebe, and D.S. Schwartz (MRS, Pittsburgh, PA, 1995), p. 793.

[60] J. Lindemann, C. Buque and F. Appel, Acta Mater. 54, 1155 (2006).

8

Mathematical Modeling of Single Peak Dynamic Recrystallization Flow Stress Curves in Metallic Alloys

R. Ebrahimi and E. Shafiei

Department of Materials Science and Engineering, School of Engineering,
Shiraz University, Shiraz,
Iran

1. Introduction

Deformation of metals and alloys at temperatures above 0.5 T_m is a complex process in which mechanical working interacts with various metallurgical processes such as dynamic restoration including recovery and recrystallization, and phase transformation for polymorphous materials. The understanding of these processes, however, enables the behavior of the metals and alloys. Recent developments have been described in several review papers. These reviews were in agreement that metals and alloys having relatively low values of stacking fault energy (SFE) could recrystallize dynamically, whereas those of high SFE including bcc metals and alloys which behave in a manner similar to fcc metals of high SFE recovered dynamically only during high temperature deformation. So that, according to microstructural evolutions, material response can principally be divided into two categories in hot deformation: dynamic recovery (DRV) type and dynamic recrystallization (DRX) type.

The process of recrystallization of plastically deformed metals and alloys is of central importance in the processing of metallic alloys for two main reasons. The first is to soften and restore the ductility of material hardened by low temperature deformation that occurring below about 50% of the absolute melting temperature which leads to lower forces. The second is to control the microstructure and mechanical properties of the final product.

The analysis of metal forming process such as hot rolling, forging and extrusion has been dependent on various parameters including constitutive relation which contributes to stress-strain curves at high temperatures, shape of workpiece and product, shapes of tools, friction, temperature, forming speed, etc. In such parameters, the constitutive equation is one of the most important factors which have an effect on solution accuracy.

A number of research groups have attempted to develop constitutive equations of materials and suggested their own formulations by putting the experimentally measured data into one single equation. Misaka and Yoshimoto proposed a model which gives the flow stress of carbon steels as a function of the strain, strain rate, temperature and carbon content. Shida's model takes account of the flow stress behavior of the steels in austenite, ferrite and in the two-phase regions. Voce suggested an approximate equation of stress–stain curve considering the dynamic recrystallization. Finally, Ebrahimi et al. obtained a mathematical model according to the phenomenological representation of the shape of the stress-strain curves that consists an additional constant. Due to the importance of flow stress estimation of metals and alloys at high temperatures, the required forces for the deformation processes, dimensional accuracy of final products and simulation of processes; it is necessary to provide a model in order to eliminate the limitations of previous models to some extent. The contexts of this section could generally be classified into two main parts, at the beginning, some previous mathematical models and basic concepts in relation with dominate processes during hot deformation of metals and alloys, such as dynamic recovery (DRV) and dynamic recrystallization (DRX) will be reviewed and finally, by concluding from existing mathematical models related to prediction of single peak flow stress curves, a new model will be introducd. With regards to the importance of macroscopic data from mechanical tests as compared to microscopic ones from metallurgical investigation, due to its less time and cost consuming nature, mathematical and macroscopic aspects of DRX process are considered in this model.

2. Basic concepts

2.1 Dynamic recovery (DRV)

The basic mechanisms of dynamic recovery are dislocation climb, cross-slip and glide, which result in the formation of low angle boundaries as also occurs during static recovery. However, the applied stress provides an additional driving pressure for the movement of low angle boundaries and those of opposite sign will be driven in opposite directions, and this stress-assisted migration of dislocation boundaries may contribute significantly to the overall strain.

Such migration results in some annihilation of dislocations in opposing boundaries and Y-junction boundary interactions and these enable the subgrains to remain approximately equiaxed during the deformation. In-situ SEM deformation experiments have shown that some reorientation of subgrains may also occur during hot deformation. The subgrains can therefore be considered to be transient microstructural features.

The processes of work hardening and recovery lead to the continual formation and dissolution of low angle boundaries and to a constant density of unbound or 'free' dislocations within the subgrains. After a strain of typically 0.5 to 1, the subgrain structure often appears to achieve a steady state. The microstructural changes occurring during dynamic recovery are summarized schematically in figure 1.

2.2 Dynamic recrystallization (DRX)

In metals which recovery processes are slow, such as those with a low or medium stacking fault energy (copper, nickel and austenitic iron), dynamic recrystallization may take place when a critical deformation condition is reached. A simplified description of the phenomenon of dynamic recrystallization is as follows. As shown in figures 2, new grains originate at the old grain boundaries, however, as the material continues to deform, the dislocation density of the new grains increases, thus reducing the driving force for further growth, and the recrystallizing grains eventually cease to grow. An additional factor which may limit the growth of the new grains is the nucleation of further grains at the migrating boundaries.

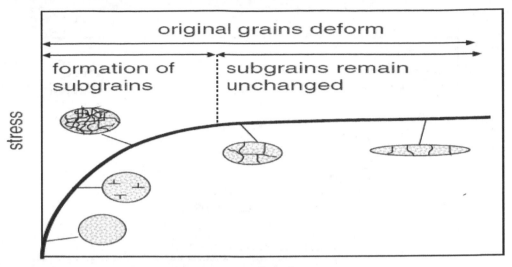

Fig. 1. Summary of the microstructural changes which occur during dynamic recovery.

This type of dynamic recrystallization, which has clear nucleation and growth stages, can be classified as a discontinuous process. There are other mechanisms which produce high angle grain boundaries during high temperature deformation and which may be considered to be types of dynamic recrystallization such as continuous dynamic recrystallixation (CDRX).

The general characteristics of dynamic recrystallization are as follows (Humphreys et al., Elsevier):

- As shown in figure 3, the stress-strain curve for a material which undergoes dynamic recrystallization generally exhibits a broad peak that is different to the plateau, characteristic of a material which undergoes only dynamic recovery (fig.1). Under conditions of low Zener–Hollomon parameter, multiple peaks may be exhibited at low strains, as seen in figure 3.
- A critical deformation is necessary in order to initiate dynamic recrystallization.
- The critical deformation decreases steadily with decreasing stress or Zener–Hollomon parameter (description in section 3.2), although at very low strain rates (creep) the critical strain may increase again.
- The size of dynamically recrystallized grains increases monotonically with decreasing stress. Grain growth does not occur and the grain size remains constant during the deformation.
- The flow stress and grain size are almost independent of the initial grain size, although the kinetics of dynamic recrystallization are accelerated in specimens with smaller initial grain sizes.
- Dynamic recrystallization is usually initiated at pre-existing grain boundaries although for very low strain rates and large initial grain sizes, intragranular nucleation becomes more important.

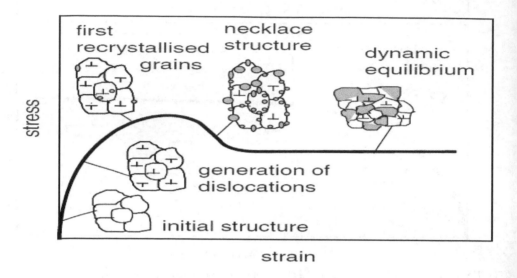

Fig. 2. Evolution of microstructure during hot deformation of a material showing DRX.

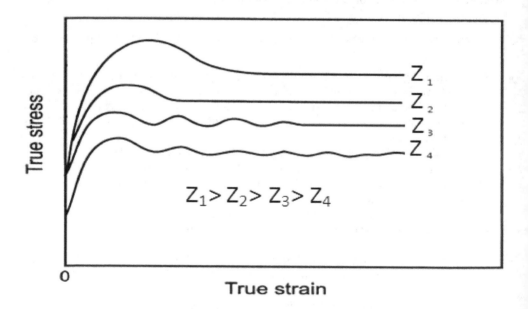

Fig. 3. The effect of Zener Hollomon parameter on the stress- strain curve.

2.3 Test methods

There are many different types of test methods, such as hot torsion, compression, tension, cam plastometer and drop hammer for developing the constitutive equation of the materials behavior. Among them, hot torsion and compression tests have widely been used. Hot torsion test has been used by many researchers to formulate constitutive equation of materials subject to a large strain because it has a forte in simulating the multi-pass deformation, in comparison with axisymmetric compression test. In the case of compression test, high friction at the interface of material and stroke head results in barreling during test. Thus, the compression test has a limitation in generating flow stress curve when the material undergoes large strain. Also Kim et al. have shown that the flow stress obtained from compression test is higher than that from torsion test, although the general shapes of the measured curves are similar. The differences between the stresses were approximately in the range of 10–20%.

These differences might be attributed to the following reasons: compression test has the glowing frictional forces at the ram-specimen interface as the test progress, while, there is no frictional effect in torsion test. In compression test, it is difficult to achieve constant strain rate and isothermal condition during test, whereas to control them in torsion test is accurate.

2.4 Flow curves

For metals with high stacking fault energy which experience DRV, the flow stress curves increase with strain in the initial deformation and reach constant in consequence of attaining the balance between work hardening and DRV (saturation stress, σ_s). For metals with DRX, initially the flow stress increases with strain which is being dominated by work hardening, and as DRX takes place upon critical strain (ε_c), the flow stress begins to decrease after reaching certain peak value. When the equilibrium is reached between softening due to DRX and work hardening, the curves drop to a steady state region (σ_{ss}). As shown in figure 3, the stress strain curves of a dynamically recrystallizing material may be characterized by a single peak or by several oscillations. Luton and Sellars have explained this in terms of the kinetics of dynamic recrystallization. At low stresses, the material recrystallizes completely before a second cycle of recrystallization begins, and this process is then repeated. The flow stress, which depends on the dislocation density, therefore oscillates with strain. At high stresses, subsequent cycles of recrystallization begin before the previous ones are finished, the material is therefore always in a partly recrystallized state after the first peak, and the stress strain curve is smoothed out, resulting in a single broad peak. Fig. 4 shows a schematic representing of dynamic recovery and a single peak dynamic recrystallization.

2.5 Strain hardening rate versus stress

The change in the slope ($\Theta=d\sigma/d\varepsilon$) of the stress-strain curve with stress can be a good indication of the microstructural changes taking place in material. All of the Θ-σ curves for a particular alloy originate from a common intercept Θ_0 at $\sigma=0$. The Θ-σ curves have three segments, two of them are linear. The first linear segment decreases with stress for initial strain to the point where subgrains begin to form with a lower rate of increase in DRV. The curve gradually changes to the lower slope of the second linear segment up to the point where the critical stress σ_c is attained for initiating the dynamic recrystallization. The curve then drops off rather sharply to $\Theta=0$ at peak stress. Extrapolation of the second linear segment of the Θ-σ curve intercepts the σ axis at the saturation stress. This would be the shape of the Θ-σ curve if dynamic recrystallization were absent with only dynamic recovery as the restorative mechanism operation, Fig.5.

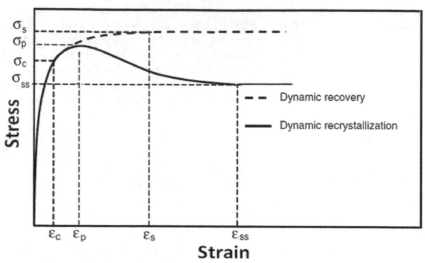

Fig. 4. Schematic representing of DRV and DRX.

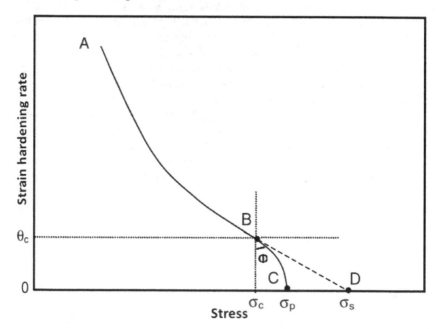

Fig. 5. Changes in the slope of the stress-strain curve with stress.

3. A review on presented models

Considerable researches have been carried out to model flow stress curves at high temperatures based on both empirical and mathematical models. Empirical models which are based on trial and error mainly obtained by repeating the mechanical tests at different conditions, and among these models, Shida and Misaka models could be mentioned. The most important limitation of these types of models is that they are valid for specific conditions and cannot be generalized, whereas in mathematical models, in order to estimate the flow curve, the material response are modeled first and then the empirical data is used to verify the prediction. Among these models, Cingara, Kim et.al and Ebrahimi et.al could be mentioned. According to the presented models, some limitations could be considered in this category, too. For instance, the presented model by McQueen and Cingara for DRX is only valid up to the peak stress (σ_p) and does not consider the softening behavior due to DRX, although a complement model has been developed to predict the softening behavior due to DRX.

Mathematical modeling of flow stress curves can be divided into three main topics as follows:

- Mathematical models which characterize the initiation of DRX.
- Mathematical models which predict characteristic points of flow stress curves.
- Mathematical models which predict flow stress curves.

It is important to mention that, due to less time and cost consuming nature of mathematical and macroscopic models, only this aspect of DRX process is considered in this chapter.

3.1 Initiation of dynamic recrystallization

The critical strain for initiation of DRX could be determined by metallography. However, this technique requires extensive sampling before and after the critical strain. Furthermore, phase changes during cooling from hot working temperature alter the deformed structure, which in turn render difficulties for metallographic analysis. Also this technique requires a large number of specimens deformed to different strains. On the other hand, the critical strain thus obtained is not precise.

Several attempts have been made to predict the initiation of DRX. For example, Ryan and McQueen observed that the presence of a stress peak at a constant strain rate flow curve leads to an inflection in plots of strain hardening rate, θ, versus stress, σ. Moreover, the points of inflection in θ-σ plots where the experimental curves separate from the extrapolated lower linear segments give critical conditions for initiation of DRX. Later, Poliak and Jonas have shown that this inflection point corresponds to the appearance of an additional thermodynamic degree of freedom in the system due to the initiation of DRX.

3.1.1 Determination of critical stress

The simple method of Najafizadeh and Jonas was used for determination of the critical stress for initiation of DRX. The inflection point is detected by fitting a third order polynomial to the θ-σ curves up to the peak point as follows:

$$\theta = A\sigma^3 + B\sigma^2 + C\sigma + D \tag{1}$$

where A, B, C, and D are constants for a given set of deformation conditions. The second derivative of this equation with respect to σ can be expressed as:

$$\frac{d^2\theta}{d\sigma^2} = 6A\sigma + 2B \tag{2}$$

At critical stress for initiation of DRX, the second derivative becomes zero. Therefore,

$$6A\sigma_c + 2B = 0 \Rightarrow \sigma_c = \frac{-B}{3A} \tag{3}$$

An example of θ-σ curves and its corresponding third order polynomial are shown in Fig. 6. Therefore, this method is used to determine the value of critical stress at different deformation conditions. Using the flow curves, the values of critical strain are determined (Najafizadeh et al. 2006).

According to Fig. 7, the normalized critical stress and strain for 17-4 PH stainless steel could be presented as:

$$\frac{\sigma_c}{\sigma_p} = 0.89 \tag{4}$$

$$\frac{\varepsilon_c}{\varepsilon_p} = 0.467 \tag{5}$$

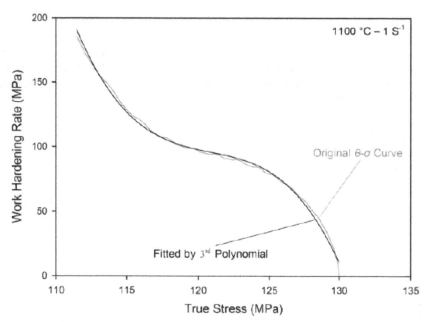

Fig. 6. The θ-σ curve of 17-4 PH Stainless steel and its corresponding third order polynomial.

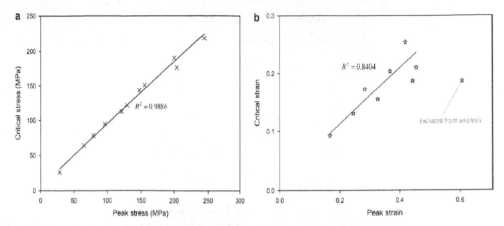

Fig. 7. Critical stress and strain versus (a) peak stress and (b) peak strain.

3.1.2 Determination of critical strain, (Poliak and Jonas)

Poliak and Jonas have shown that the inflection in plots of $\ln\Theta$-ε and $\ln\Theta$-σ curves can also be used for determination of initiation of DRX. The procedure of Section 3.1.1 is used to determine the values of critical strain at different deformation conditions (Mirzadeh et al., 2010).

According to figure 8, the normalized critical strain could be presented for 17-4 PH stainless steel:

$$\frac{\varepsilon_c}{\varepsilon_p} = 0.47 \qquad (6)$$

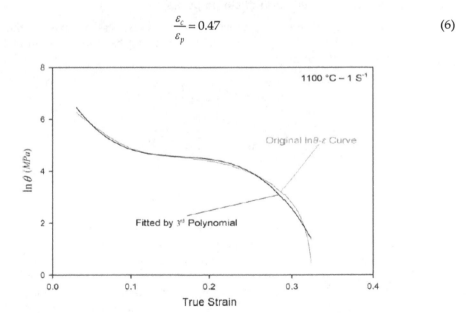

Fig. 8. The $\ln\Theta$-ε curve for 17-4 PH stainless steel and its corresponding third order polynomial.

3.1.3 Determination of critical strain (Ebrahimi and Solhjoo)

The flow stress up to the peak stress was modeled using the Cingara and McQueen equation as shown below:

$$\sigma = \sigma_p[(\varepsilon / \varepsilon_p)\exp(1 - \varepsilon / \varepsilon_p)]^C \tag{7}$$

Where constant C is an additional parameter to make the results acceptable and is obtained from logarithmic form of Eq.7. The derivative of the true stress with respect to true strain yields the work hardening rate, Θ. Therefore, the Θ formula using Eq.7 may be expressed as:

$$\theta = C\sigma(1 / \varepsilon - 1 / \varepsilon_p) \tag{8}$$

In order to determine the critical strain, the second derivative of Θ with respect to σ must be zero. By solving this equation, the critical strain as a function of peak strain will obtain as follows:

$$\frac{\varepsilon_c}{\varepsilon_p} = \frac{\sqrt{1 - C} - (1 - C)}{C} \tag{9}$$

The results showed a good agreement with the experimentally measured ones for Nb-microallyed steel (Ebrahimi et al., 2007).

3.2 Kinetic equations (prediction of single points)

The influence of temperature and strain rate on peak stress was analyzed by the following equations which were originally developed for creep but have found applicability in the high strain rates in hot working:

$$A'\sigma_p{}^{n'} = \dot{\varepsilon}\exp\left(\frac{Q_{HW}}{RT}\right) = Z \tag{10}$$

$$A''\exp(\beta\sigma_p) = \dot{\varepsilon}\exp\left(\frac{Q_{HW}}{RT}\right) = Z \tag{11}$$

$$A'''\left[\sinh(\alpha\sigma_p)\right]^{n} = \dot{\varepsilon}\exp\left(\frac{Q_{HW}}{RT}\right) = Z \tag{12}$$

where $A', A'', A''', n, n'', Q_{HW}, \alpha, \beta$ and R are constants and Z is the Zener Hollomon. Q_{HW} is the activation energy related to hot working and R is the universal gas constant. A power law plot of log σ versus log Z gives linear segments only at low stress, indicating the limited applicability of Eq. (10). However in a plot of σ against log Z linearity is lost at low stress, showing that Eq. (11) is an inadequate fit for the entire range of stresses. Eq. (12) is a more general form of Eq. (10) and Eq. (11) reducing to Eq. (10) at lower stresses ($\alpha\sigma < 0.8$) and Eq. (11) at higher stresses ($\alpha\sigma > 1.2$). The validation of these equations to predict characteristic points of flow curves were approved by many authors (Cingara et al., 1992).

3.3 Experimental models

3.3.1 Misaka's model

Misaka and Yoshimoto have utilized the following double-power constitutive equation to determine flow stress associated with the processing of steels:

$$\sigma_{Misaka} = 9.8\exp\left(0126 - 1.75[C] + 0.594[C]^2\right) + \left(\frac{2851 + 2968[C] - 1120[C]^2}{T + 273}\right)\varepsilon^n \dot{\varepsilon}^m \qquad (13)$$

Application range of this formula is as follows; carbon content: ≈1.2%, temperature: 750–1200 °C, reduction (natural strain): 50% and strain rate: 30–200 s^{-1}. Misaka's equation was updated by including effects of solution-strengthening and dynamic recrystallization.

The updated Misaka's constitutive equation is:

$$\sigma_{Misaka,Updated} = \left(f\sigma_{Misaka}\right)\left(1 - X_{DRX}\right) + k\sigma_{ss}X_{DRX} \qquad (14)$$

$$f = 0.835 + 0.51[Nb] + 0.098[Mn] + 0.128[Cr]^{0.8} + 0.144[Mo]^{0.3} + 0.175[V] + 0.01[Ni] \quad (15)$$

(σ_{Misaka})updated indicates the flow stress of steels containing multiple alloying additions. σ_{ss} is steady state stress, $K = 1.14$ is a parameter that converts flow stress to mean flow stress, and X_{DRX} is volume fraction of dynamic recrystallization. It might be useful for practical purpose but its mathematical base is weak. The factors for Mn, Nb, V and Ni are linear although the terms for Cr and Mo are nonlinear. Devadas et al. compared the predicted flow stress data for a low alloy steel with measured data from a cam-plastometer. They showed that Misaka's model overestimated the flow stress (Kim et al., 2003).

3.3.2 Shida's equation

Shida's equation is based on experimental data obtained from compression type of high strain rate testing machines. Shida then expressed the flow stress of steels, σ, as a function of the equivalent carbon content (C), the strain (ε), the strain rate ($\dot{\varepsilon}$) and temperature (T) as followings:

$$\sigma = \sigma_d(C,T)f_w(\varepsilon)f_r\left(\dot{\varepsilon}\right) \qquad (16)$$

$$\sigma_d = 0.28\exp\left(\frac{5}{T} - \frac{0.01}{C + 0.05}\right) \qquad (17)$$

$$T[k] = \frac{T[^\circ C] + 273}{1000} \qquad (18)$$

$$f_w(\varepsilon) = 1.3\left(\frac{\varepsilon}{0.2}\right)^n - 0.3\left(\frac{\varepsilon}{0.2}\right) \qquad (19)$$

$$n = 0.41 - 0.07C \tag{20}$$

$$f_r\left(\dot{\varepsilon}\right) = \left(\frac{\dot{\varepsilon}}{10}\right)^m \tag{21}$$

$$m = \left(-0.019C + 0.126\right)T + \left(0.076C - 0.05\right) \tag{22}$$

where $f_w(strain)$ and $f_r(strain\ rate)$ are functions dependent upon strain and strain rate, respectively. The formulation of Eq. (16) is based on assumption that flow stress increases with the strain rate and strain increased. The range of validity of the formula is quite broad. This formula is applicable in the range of carbon content: 0.07–1.2%, temperature: 700–1200°C, strain: ≈ 0.7 and strain rate: $\approx 100\ s^{-1}$.

3.3.3 Modified Voce's equation

In contrast to Misaka's and Shida's equations, Voce's equation can describe the flow stress over the wide range of strains and strain rates. The equation can express the dynamic softening portion of the flow stress curve by using Avrami equation:

$$\sigma_{WH+DRV} = \sigma_p\left[1 - \exp(-c\varepsilon)\right]^m \tag{23}$$

The coefficient, C, and work hardening exponent, m, are dependent on the deformation conditions. The parameters, C and m are normally taken as being a constant, however, it is a function of the deformation conditions (strain rate, temperature). Thus, the C and m are considered to be a function of dimensionless parameter, Z/A.

Also Kim et al. developed an equation by modifying Voce's constitutive equations accounting for the dynamic recrystallization as well as the dynamic softening. During thermomechnical processing, the important metallurgical phenomena are work hardening (WH), dynamic recovery (DRV) and dynamic recrystallization (DRX). Thus, the flow stress curve can be bisected with the WH + DRV region and the DRX region. For the evaluation of the WH and DRV region, Eq. 23 was used. Also For the region of DRX, the drop of flow stress was expressed as the following equation:

$$\sigma_{Drop} = \left(\sigma_p - \sigma_{ss}\right)\left[\frac{X_{DRX} - X_{\varepsilon_p}}{1 - X_{\varepsilon_p}}\right] \quad \text{for } \varepsilon > \varepsilon_p \tag{24}$$

Therefore, the flow stress can be expressed in subtraction form as follows:

$$\sigma = \sigma_{WH+DRV} - \sigma_{Drop} \tag{25}$$

where σ_{ss} is the steady state stress achieved at larger strains and $X_{\varepsilon p}$ the volume fraction of DRX at peak strain. X_{DRX} is the volume fraction of DRX at any strain.

Fig. 9 shows the measured and predicted flow stress curves in large strain range of AISI 4140 steel when the three different types of constitutive equations are used for prediction. The flow stress curves calculated by using Misaka's equation agree with the measured ones with some extent of error in comparison to the flow stress curve obtained from Shida's equation. Although, the stress–strain curves predicted by using the modified Voce's equation are in a good agreement with experimentally measured ones, it seems Misaka's equation does not reflect recrystallization behavior properly (Kim et al., 2003).

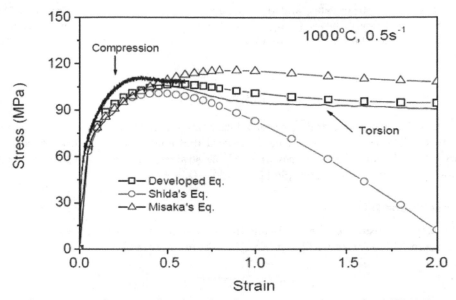

Fig. 9. Comparison of measured and predicted constitutive relations for AISI 4140 steel.

3.4 Mathematical models

3.4.1 Solhjoo's model

Solhjoo was considered linear estimation of θ-σ curve up to the peak stress as follows:

$$\frac{\partial \sigma}{\partial \varepsilon} = S_4 + A_1 \tag{26}$$

where S_4 is the slope of the line and A_1 is a constant. Due to the first assumption which considers a linear segment up to the peak, this model would not be able to show the critical strain. Using the maximum point of the θ-σ curve ($\sigma=\sigma_p, \theta=0$), value of constant A_1 is obtained to be $-S_4\sigma_p$. Solution of the differential Eq. (26) with respect to ε, using boundary condition $\sigma=\sigma_p$ at $\varepsilon=\varepsilon_p$ is:

$$\sigma = \sigma_p \left[1 - \left(\frac{(\varepsilon_p - \varepsilon)}{\varepsilon_p} \right) \exp(S_4 \varepsilon) \right] \tag{27}$$

where σ_p is the peak stress and ε_p is the peak strain.

Determination of S_4 can be simply done using a linear plot of $\ln\left(\dfrac{\varepsilon_p}{\sigma_p}\dfrac{\sigma_p-\sigma}{\varepsilon_p-\varepsilon}\right)$ vs. ε; S_4 would

be the slope of the line. As S_4 is very sensitive to strain rate, it is better to plot $\ln\left(\dfrac{\varepsilon_p}{\sigma_p}\dfrac{\sigma_p-\sigma}{\varepsilon_p-\varepsilon}\right)$
vs. ε for each sets of strain rates, and then the average value of the slopes determines S_4.
Assuming S_4 as a constant shows a rough estimation of stress– strain curve. Since S_4 is more
sensitive to strain rate than temperature, this parameter can be considered as a power law in
form of:

$$S_4 = -C\dot{\varepsilon}^E$$

(28)

where E is strain rate and C and ε are constants. Using a plot of $\ln(-S_4)$ vs. $\dot{\varepsilon}$, constants C
and E can be determined. It should be considered that another limit of this model is its
disability of prediction of flow stress at very low strains (less than 0.05) that the work
hardening rate has very high values (Solhjoo, 2009).

3.4.2 Avrami's analysis

The DRX may be considered as a solid-state transformation and its kinetics can be modeled
by the Johnson–Mehl–Avrami–Kolmogorov (JMAK) equation as follows:

$$X_{DRX} = 1 - \exp\left(-kt^n\right)$$

(29)

where X_{DRX} and t are the recrystallized volume fraction and DRX time, respectively. The
effect of dynamic recovery (DRV) on flow softening was not considered. Therefore, this
model is preliminary intended for materials with relatively low stacking fault energy.
Moreover, for modeling the flow curves after the peak point of stress–strain curve, the
initiation of DRX was intentionally considered at peak point. This assumption simplifies the
Avrami analysis with acceptable level of accuracy. Therefore, Eq. (30) gives the magnitude
of flow stress at each fractional softening:

$$\sigma = \sigma_p - \left(\sigma_p - \sigma_{ss}\right)X_{DRX}$$

(30)

3.4.3 Ebrahimi's model

This model is based on a phenomenological representation of the shape of the flow stress
curves and the traditional theories for constitutive equations which incorporate the power
law. Ebrahimi et al. considered the variations of the slope of flow stress curves as follows:

$$\frac{d\sigma}{d\varepsilon} = C_1\left(\sigma - \sigma_{ss}\right)\left(1 - \frac{\varepsilon}{\varepsilon_p}\right)$$

(31)

Where ε_p is the strain at the peak stress and σ_{ss} is the steady state stress. The term $(1-\varepsilon/\varepsilon_p)$ estimates variation of the stress- strain curve for $\sigma > \sigma_{ss}$. Solution of the differential Eq.(31) with respect to ε using boundary condition $\sigma_{=}\sigma_p$ at $\varepsilon_{=}\varepsilon_p$ results in:

$$\sigma = \sigma_{ss} + \left(\sigma_p - \sigma_{ss}\right)\exp\left[C_1\left(\varepsilon - \frac{\varepsilon_p}{2} - \frac{\varepsilon^2}{2\varepsilon_p}\right)\right] \tag{32}$$

If $\varepsilon_{=}\varepsilon_k$ and $\varepsilon_{k=}k\varepsilon_p$ where $k<1$ and $\sigma_k>\sigma_{ss}$, then coefficient C_1 can be estimated from Eq. (33) as:

$$C_1 = \frac{2}{\left(k^2 - 2k + 1\right)\varepsilon_p}\ln\left(\frac{\sigma_p - \sigma_{ss}}{\sigma_k - \sigma_{ss}}\right) \tag{33}$$

Where σ_k is the stress calculated from Cingara equation. The equation represented by this model required the values of stress and strain at the peak and stress at the steady state zone. These parameters can be calculated by kinetic equations (Ebrahimi et al., 2006).

Fig. 10 shows the calculated flow curves by Cingara, Avrami and Ebrahimi et al. equations. At low Z, the predictions of Ebrahimi et al. equation are relatively accurate (Fig. 10a), but at high Z, this equation overestimates the flow softening of DRX (Fig. 10c). In other words, this equation seems to be suitable for ideal DRX behavior. The Ebrahimi et al. equation is based on a phenomenological representation of the shape of the stress–strain curves and the traditional theories for constitutive equations which incorporate the power law. Conversely, the Avrami equation is amenable to all deformation conditions as shown in Fig. 10. In fact, this equation is an adaptive one with two adjustable constants. As a result, it could be better fitted to hot flow curves. Therefore, the Avrami equation can be used for prediction of flow curves as shown in Fig. 10 (Mirzadeh et al., 2010).

3.4.4 Shafiei and Ebrahimi's constitutive equation

Using the extrapolation of DRV flow stress curves and kinetic equation for DRX, Shafiei and Ebrahimi proposed the following equation for modeling single peak DRX flow curves for $\varepsilon_c \leq \varepsilon < \varepsilon_{ss}$

$$\sigma = \sigma_s - \left(\sigma_s - \sigma_c\right)\exp(C'\left(\varepsilon_c - \varepsilon\right)) - \left(\sigma_s - \sigma_{ss}\right)X_{DRX} \tag{34}$$

Where C' is a constant with metallurgical sense. According to the geometrical relations shown in Fig.5, the value of C' can be formulized as Eq.s (35) and (36) . σ_c, σ_p, σ_{ss}, σ_s , ε_c , X_{DRX} are critical stress, peak stress, steady state stress, saturation stress , critical strain and volume fraction of DRX, respectively.

$$\tan\varphi = \frac{1}{C'} \tag{35}$$

$$\varphi = \frac{\pi}{2} + Arctg\left(\frac{\sigma_p \varepsilon_p}{\varepsilon_c(\varepsilon_c - \varepsilon_p)}\right) \tag{36}$$

As shown in Fig.11, the stress-strain curves predicted by using presented model are in a good agreement with experimentally measured ones for Ti-IF steel. In order to evaluate the accuracy of the model, the mean error was calculated. The mean error of flow stress is calculated at strains of 0.19-2 for every measurement under all deformation conditions. For all stress-strain curves, the mean errors are between -2.9% and 2.5%. The results indicate that the proposed model give a good estimate of the flow stress curves. Therefore, it can be deduced that the approach to obtain a constitutive equation applicable to large strain ranges was fruitful and this presented model might have a potential to be used where more precise calculation of stress decrement due to dynamic recrystallization is important. Moreover, this analysis has been done for the stress-strain curves under hot working condition for Ti-IF steel, but it is not dependent on the type of material and can be extended for any condition that a single peak dynamic recrystallization occurs.

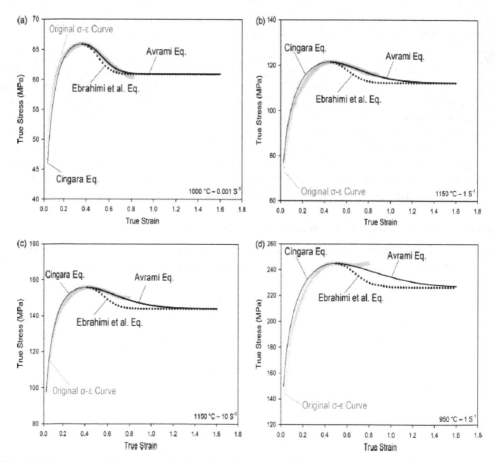

Fig. 10. Comparison between calculated and measured flow stress curves of 17-4 PH stainless steel.

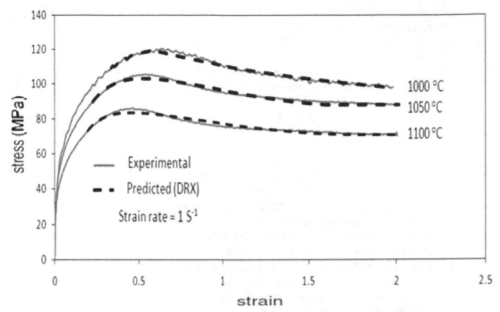

Fig. 11. Comparison of measured and predicted stress-strain curves of Ti-IF steel.

4. Summary

In this chapter, a review on recent model of single peak flow stress curves was presented. At first, the basic concepts on hot deformation and dynamic restoration, including affecting factors on dominated processes such as DRX and DRV and related microstructure evolutions were discussed. Then, an introduction on experimental models which are more capable in the field of industrials investigations such as Misaka, Shida and Voce constitutive equations followed by the details of mathematical models such as MC Queen, Ebrahimi, Solhjoo, Avrami and Shafiei- Ebrahimi constitutive equations were presented. In this case, the accuracy of these models as well as some limitations was evaluated in order to obtain the optimum working conditions.

5. References

Cingara, A. & McQueen, H. J. (1992). New method for determining sinh constitutive constants for high temperature deformation of 300 austenitic steel. Mater. Proc. And Tech, 36, 17-30.

Ebrahimi, R. & Solhjoo, S. (2007). Characteristic points of stress-strain curve at high temperature. ISSI, 2, 24-27.

Ebrahimi, R., Zahiri, S.H. & Najafizadeh, A. (2006). Mathematical modeling of the stress-strain curves of TI-IF steel at high temperature. Mater. Proc. And Tech., 171, 301-305.

Ebrahimi, R. (2003). Hot working of Ti-IF steel. Ph.D thesis. Isfahan University of Technology.

He, X. Yu, Zh. & Lai, X. (2008). A method to predict flow stress considering dynamic recrystallization during hot deformation. Comp. Mater. Sci., 44, 760-764.

Humphreys, F.J. & Hatherly, M. (2004). *Recrystallization and related annealing phenomena* (Second edition), (Elsevier, UK).

Imbert, C. A. C & McQueen, H. J. (2001). Peak strength, strain hardening and dynamic restoration of A2 and M2 tool steels in hot deformation. Mater. Sci. and Eng. A, 313, 88-103.

Imbert, C. A. C & McQueen, H. J. (2001). Dynamic recrystallization of A2 and M2 tool steels. Mater. Sci. and Eng. A, 313, 104-116.

Jonas, J. J., Quelennec, Jiang, L. & Martin, E. (2009). The avrami kinetics of dynamic recrystallization. Acta Mater. 137, 1748-2756.

Kim, S. I., Lee, Y. & Byon, S. M. (2003). Study on constitutive relation of AISI 4140 steel subject to large strain at elevated temperature. Mater. Proc. And Tech.,140, 84-89.

Liu, J., Cai, Zh. & Li, C. (2008). Modelling of flow stress characterizing dynamic recrystallization for magnesium alloys. Comp. Mater. Sci., 41,375-382.

Mirzadeh, H. & Najafizadeh, A. (2010). Extrapolation of flow curves at hot working conditions. Mater. Sci. and Eng. A, 527,1856-1860.

Mirzadeh, H. & Najafizadeh, A. (2010). Prediction of the critical conditions for initiation of dynamic recrystallization. Mater. And Des., 31, 1174-1179.

Najafizadeh, A. & Jonas, J. J. (2006). Predicting the critical stress for initiation of dynamic recrystallization. ISIJ Int., 46, 1679-1684.

Poliak, E. I. & Jonas, J. J. (2002). Initiation of dynamic recrystallization in constant strain rate hot deformation. ISIJ Int., 43, 684-691.

Shaban, M. & Eghbali, B. (2010). Determination of critical conditions for dynamic recrystallization of a microalloyed steel. Mater. Sci. and Eng. A, 527, 4320-4325.

Shafiei, E. & Ebrahimi, R. (2011). Mathematical modeling of single peak flow stress curves. Comp. Mater. Sci., submitted paper.

Solhjoo, S. (2009). Analysis of flow stress up to the peak at hot deformation. Mater. And Des.,30, 3036-3040.

Ueki, M., Horie, S. & Nakamura, T. (1987). Factors affecting dynamic recrystallization of metals and alloys. Mater. Sci. and Tech.,3, 329.

Verlinden, B., Driver, J., Samijdar, I. & Doherty, R. D. (2007). Thermomechanical processing of metallic materials. (Elsevier, Uk).

Zahiri, S. H., Davies, C. H. J. & Hodgson, P. D. (2005). A mechanical approach to quantify dynamic recrystallization in polycrystalline metals. Scripta Mater., 52, 299-304.

Zeng, Zh., Jonsson, S., Rovan, H. J. & Zhang, Y. (2009). Modelling the flow stress for single peak dynamic recrystallization. Mater. And Des.,30-1939-1943.

9

The Effect of Strain Path on the Microstructure and Mechanical Properties in Cu Processed by COT Method

Kinga Rodak

Silesian University of Technology Katowice,
Poland

1. Introduction

In recent years a lot of place in research has been devoted to the methods of grain refinement with the aid of large plastic deformations- *Severe Plastic Deformation*. The research of the methods of grain refinement is carried out in parallel to the intensive research of the structural changes occurring in the deformed materials. Some metallic materials which are deformed by means of *Severe Plastic Deformation* are characterized by ultrafine-grained and sometimes even nanograined size. Grain refinement fosters, above all, the increase in the strength of a material. Therefore, a production of nano- and ultafine-grained structures has become one of the most important research issues which are currently the subject of interest in many research facilities. There are many techniques of SPD, for example: *Equal Chanel Angular Pressing* (ECAP), *High Pressure Torsion* (HPT), *Cyclic Extrusion Compression* (CEC), *Hydroextrusion* (HE), the KOBO methods, *Acumulative Roll Bonding* (ARB). The SPD techniques cannot be described as easy ways of obtaining the refined materials mainly because of the insufficient homogeneity of the structure, the low efficiency of the applied methods, and substantial losses of the material [Pakieła, 2009; Olejnik, 2005; Cao 2008]. That is why, the commonly know methods of SPD, excluding some exceptions [Bochniak, 2005], have not become implemented as production technologies, although a lot of time has passed since they were discovered. This is the reason why more and more works concentrate on using SPD techniques to produce a structure that is more homogenous and has higher degree of refinement [Shaarbaf, 2008; Raab, 2004]. There also appear some suggestions to modify SPD techniques [Lugo, 2008; Kulczyk, 2007]. Some intensive research is carried out aiming at preparing new SPD techniques which would have refined a grain to the level of ultrafine-grained or even nano-grained. One of such methods is the Compression with Oscillatory Torsion (COT). The method is regarded as an unconventional way of volume shaping and is developed by the Department Of Materials Technology in the Faculty of Materials Engineering and Metallurgy of the Silesian University of Technology. This method has become recognized mainly as a method that enables deformation of the materials to values of large plastic deformations [Pawlicki, 2007], therefore, it is possible to obtain a refined structure. The benefits from applying the COT method are visible mainly in the aspect of reduction of work hardening effects (lowering of the plastic deformation work) [Grosman, 2006] and formation of a particular type of a spatial configuration of defects in deformed material [Rodak, 2007]. It is, therefore, worth-exploring because of the attractive perspectives of using this method to form ultrafine-grained structures by using a suitable combination of deformation parameters (change of deformation path). It was proven that grain refinement in the COT method happens when the deformation parameters are suitably chosen. Among the parameters there are: torsional frequency f, compression velocity v, absolute strain Δh, and torsion angle a. An interesting observation coming from the longstanding research on the changes of the structures that accompany the COT torsion, is the fact that the effective deformation ε_f cumulated in a material is not the most important parameter thanks to which obtaining the suitably refined grains is possible.

2. Compression with oscillatory torsion (COT) method and deformation path

Compression with oscillatory torsion is a new method of plastic deformation in which the material is deformed as an effect of a changing deformation path. Fig. 1 is a schematic presentation of the COT set up. The facility for compression with oscillatory torsion consists of upper and lower anvils made from high-strength tool steel. Torsional straining was achieved by rotating the lower anvil, and compression was simultaneously achieved by linear strain from the lower anvil.

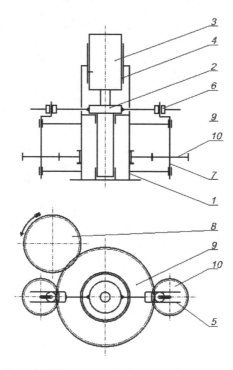

Fig. 1. Schematic illustration of COT process: 1. Frame, 2. Lower punch, 3. Upper punch, 4. Non-rotating slidable bearing, 5. Lower punch arm, 6. Roller, 7. Crankshaft, 8. Driving gear, 9. Ring gear, 10. Gear wheel

The appliance allows for the following parameters to change:

- the compression velocity v, (the velocity of the lower punch shift). The maximal value of compression velocity is 0,6 [mm/s]
- the torsional frequency f. The frequency of the lower punch oscillation is regulated by the inverter ranging from 0 Hz to 1,8 [Hz].
- the torsional angle amplitude $\alpha[°]$. The set points of the kinematic magnitudes enable the change of the torsional angle ranging from $0°$ to $6°$
- the absolute strain Δh [mm]
- the compression force F [kN]

The compressive force F and the deformation path are registered by computer. The process can be carried out only at room temperature.

The total cumulated value of deformation equals the sum of deformation dimensions obtained in consecutive cycles of torsion with simultaneous compression, for the following parameters: torsional velocity v, reversal torsional frequency f, amplitudes of the torsional angle α, the initial height of the test piece h_0, the diameter of a test piece d_0.

The total equivalent deformation ε_f is [Grosman, 2006]:

$$\varepsilon_f = \varepsilon_h + \varepsilon_t \tag{1}$$

where:

ε_h – the effective deformation from compression
ε_t - the effective deformation from torsion

This method regarded as an unconventional way of forming, has become recognized above all as an effective way to lower the force of plastic forming of a material. The force characteristics of the process depend on the component deformations induced by the torsion and compression. The value of the component deformations induced by the torsion is dependent on: torsional angle a and torsional frequency f. The component deformations induced by the compression are dependent on the compression velocity v. The proportions of the particular components describe the deformation path d_ε [Grosman, 2006].

For the constant value of the torsional angle amplitude (a = const) the deformation path is proportional to the value of the torsional frequency to compression velocity ratio.

The force characteristics $F = f(\Delta h)$ indicates the raise of the torsional frequency f, accompanied by the constant values of the other process parameters: the absolute strain Δh, the compression velocity v and the torsional angle a, has an impact on the decrease of the axial (compression) force level (Fig.2). The course of curves shows that the double raise of the torsional frequency influences the lowering of the compression force more than 1,5 times and, additionally, reduces the work hardening. Higher torsional frequencies, in connection with the increased compression velocities, have no greater influence on the change of the average unit pressure.

In the method of compression with oscillatory torsion, a various combinations of the parameters can be used in order to obtain comparable values of the effective deformations.

The table 1. shows an example set of the deformation parameters needed to obtain similar values of the total effective deformation ε_f ~5 and ~14. For the constant values of the $\Delta h = 3$ mm and the torsional angle $\alpha=6°$, the proportions between the compression velocity v and the torsional frequency f were changed. The effective deformation ε_f and the axial forces of compression are close to the test carried on with $f = 0,1$ Hz, $v=0,015$ mm/s and $f = 0,8$ Hz, $v=0,04$ mm/s. Likewise, for the test with $f = 0,2$ Hz, $v=0,015$ mm/s and $f = 1,6$ Hz, $v=0,04$ mm/s similar values of an effective deformation ε_f and the axial forces of compression are obtained. Yet, with such parameters higher effective deformations and lower axial forces of compression are achieved.

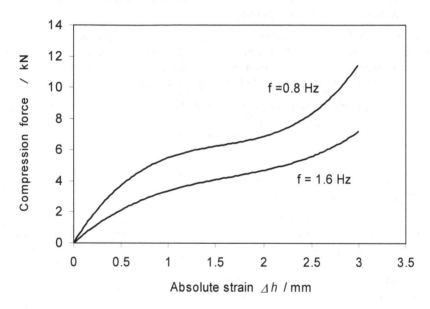

Fig. 2. Dependence of F=f(Δh) for samples 3 and 4

Sample	1	2	3	4
Parameters	$\Delta h = 3$ mm $\alpha=6°$ f= 0.1 Hz v=0.015 mm/s	$\Delta h = 3$ mm $\alpha=6°$ f= 0.2 Hz v=0.015 mm/s	$\Delta h = 3$ mm $\alpha=6°$ f= 0.8 Hz v=0.04 mm/s	$\Delta h = 3$ mm $\alpha=6°$ f= 1.6 Hz v=0.04 mm/s
Effective strain, ε_f	~5	~14	~5	~14

Table 1. Deformation parameters of Cu

The structural effects produced by the deformation with such parameters are interesting (Fig.3). Because of the possibility of using the COT method to refine the grain, more effective process is the one of deformation implemented by the use of higher parameters f and v, even for the comparable values of the effective deformations and the values of the axial forces of compression.

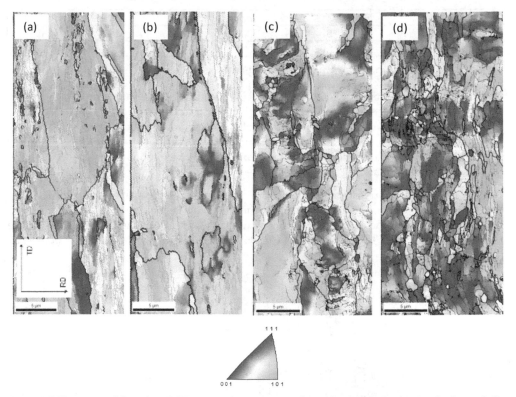

Fig. 3. EBSD maps of Cu after COT processing: a) sample 1, b) sample 2, c) sample 3, and d) sample 4

The deformation realized using high torsional frequencies evoked a decrease of the axial forces of compression in the material (Fig.2.) and the desired structural effect in a form of the grain refining (Fig.3). A similar correctness in a form of structural effects, was not observed in the case of deformable samples with small oscillating frequencies, in spite of the fact that also in this case a comparable decrease of the axial force level was registered. The conclusion might be that the effect of the decrease of the axis force and the increase of the effective deformation is not adequate with the structural effect in a form of grain refining.

In the COT method only the „special" conditions of the process (the proportions of the deformation parameters) can guarantee the effective grain refining. Therefore, the selection of the appropriate deformation parameters guarantee the effective grain refining.

3. Experimental details

The tests were conducted on the samples from copper in the M1E species (chemical composition is shown in the Table.2). This material is eagerly used to refine grains using the SPD techniques therefore, in case of new SPD techniques can constitute a good comparative material especially when it comes to the effectiveness of grain refining as well as to the mechanisms of grain refining process.

Chemical composition [%]						
Cu	Fe	Bi	Pb	Ni	Sn	As
99,8	0,02	0,002	0,008	0,019	0,03	0,001

Table 2. Chemical composition Cu used in experimental

The samples for the tests were taken from the bars having 12 mm in diameter and then, they were exposed to the heat treatment which involves the annealing in temperature of 500°C /2 hours. After this treatment the average diameter of the grain equaled 50 μm. The heating treatment that was carried out, allowed for eliminating structural effects resulting from the previous technological treatments and for obtaining the homogenous grain structure in the whole volume of the material.

The test of compression with oscillatory torsion, required two types of samples. The first series of samples was prepared according to the scheme shown in the Fig. 4, where the ratio of height h (6 mm) to the diameter d (6 mm) equaled (h/d) =1. Whereas, in the second set the height h was increased to 9 mm without the change of diameter and this equaled: (h/d) =1,5.

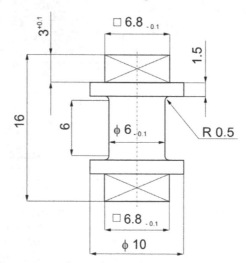

Fig. 4. Geometry and dimensions of samples used in the experiment

The deformation by means of the COT method was conducted using the following parameters:

- the range of the compression velocity; 4 variants of velocity were used: 0,015 mm/s; 0,04 mm/s; 0,1mm/s, 0,6 mm/s.

- the constant values of the torsional angle α=6°.

- the range of torsional frequency; 4 variants of this parameter were used: 0,2 Hz, 0,8 Hz, 1,6 Hz and 1,8 Hz.

- the absolute strain was Δh = 3 mm and Δh = 7 mm correspondingly for (h/d) =1 and (h/d) =1,5.

- the compression force F equaled 300 kN.

The compression with oscillatory torsion method is regarded as a method characterized by the heterogeneity of deformation. The most intense deformations proceed in places that are the nearest to the lateral surfaces of the material, which is results from the functioning of the torsional moment. The heterogeneity of the plastic deformation in the sample, causes the occurrence of a considerable differentiation of the structure in its sectional view. Because of the heterogeneity of the deformation, the microscope observations and strength testing were carried out in spheres located in a distance of about 0,8 of the sample's ray (Fig.5).

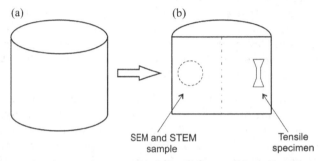

Fig. 5. Simple schematic of a specimen without the collar and handle: (a) sample after processing and (b) procedures for different types of testing: SEM and STEM observations and tensile testing using miniature specimens. The structure and mechanical properties studies were conducted on samples extracted from a distance near 0,8 radius in the longitudinal plane section

The analysis of the dislocating structure was carried out using the Scanning Transmission Electron Microscopy (STEM) technique which was applied thanks to the microscope Hitachi HD 2100A equipped with the FEG gun, working at the accelerating voltage of 200 kV.

With the help of Transmission Electron Microscopy (TEM) Jeol 100B, the orientation of the grains were determined based on the received pictures of Kikuchi lines. For the calculations the KILIN programme was used that was developed on the University of Science and Technology in Cracow.

The detailed quantities research of the ultrafine-grained structures being formed was conducted using Scanning Electron Microscope (SEM) INSPECT F produced by FEI equipped with the gun with cold field emission and the detector of electron back scattering diffraction (EBSD). In order to release the structure of the material by using the SEM/EBSD method, firstly, the mechanical polishing was used and then, electrolytic.

On the basis of the SEM/EBSD method the average equivalent diameter of the subgrains d [μm] and the average equivalent diameter of the grains D [μm] were determined. The boundary between the grain and subgrain was determined on the basis of the misorientation angle measurement. The divisional boundary was an angle equaling 15°. In the measurements of the average diameter of grain/subgrain, the grains located in the periphery of the image were not considered.

Some measurements of misorientation angles between the neighboring subgrains/grains were also taken. Therefore, the fields that had the misorientation degree from 2° were analyzed.

The estimation of the dislocation density in the material that were strongly deformed by means of transmission electron microscopy is almost impossible when these dimensions exceed $5 \cdot 10^{14}$ m^{-2}, and the result have a doubtful statistic value because of the small area of the analysis. Those inconveniences can be avoided by using the X-ray structural analysis.

The analysis of the diffraction line profile is an effective tool needed to characterize the structures of the material which have defects in crystal lattice. The presence in the material of tiny crystallites and the lattice distortions, causes the widening of diffraction reflexes. Both the size of the crystallites and the lattice distortions can be determined in quantity by means of the Williamson-Hall method in which the basis of analysis in this method is the Full Width at Half Maximum (FWHM) of the reflex, and using the Warren-Averbach method based on analyzing the Fourier coefficient [Ungar, 2001].

The difficulties in analysis using these methods occur when the material consists of anisotropy of diffraction line broadening. In such a case, neither the half-value line nor the Fourier coefficient constitute the monotonic function of the diffraction vector $K = (2sin\theta)/\lambda$, where θ - reflection angle, λ - X-ray wavelength. The replacement of K and K^2 in a classic procedure of Williamson-Hall and Warren-Averbach by the expression $K^{1/2}$ and K^2 allows for eliminating the influence of anisotropy of diffraction line broadening.

The measurements of the hardness were taken by the Vickers method using the hardness testing machine FM 700 produced by the Japanese company Future Tech coupled with the metallographical microscope. The test was carried out with the loading of HV0,2.

Applying the hardness measurement, when the small volumes of the material, usually with heterogeneous deformation are obtained in the laboratory tests, is fully justified. However, the method that describes the strength and plastic features is the static tensile test. In the case of a very small dimension of the sample obtained by the SPD methods, the tensile test can be conducted only on the micro samples. Using micro samples is connected with numerous technical inconveniences among which there are: the danger of implementing structural changes resulting from the preparation of the material for the tests, fastening of the samples and the axiality of the meter circuit. In the tests that are described here, the micro samples have the dimensions 1.20 x 0.3 and length 2.20. The cutting was made using the spark wire (the wire was 0,1mm thick) with an intense cooling in distilling water.

The tensile test was carried out by employing the universal testing machine with the screw drive equaling MTS QTest/10. This method is successfully introduced in measuring deformations of small samples by the Faculty of Materials Science and Engineering in the Warsaw University of Technology. The elongation of the samples is measured using the digital image correlation method. The method is based on comparing of the digital recording of the sample image before the deformation (the reference image) with the digital recording of the sample image after the deformation. The computer algorithm compares the images and describes the relocate of small areas within the tested surface. Thanks to this method the local and total deformations on the whole analyzed surface of the sample can be determined. The tensile test was conducted with the initial tension velocity equaling $2*10^{-3}$ [1/sec].

4. The impact of the deformation parameters on the grain refining

4.1 The torsional frequency

Among the several changing parameters of the deformation used in the COT method, the torsional frequency is the factor which, when increased, has a beneficial influence on lowering the value of the compression force (Fig.2.). Moreover, the increase of this parameter causes that the material is deformed to large values of effective deformations. For example, when the torsional frequency is 0,2 Hz and 1,8 Hz with the constant parameters: $\Delta h=7$, $\alpha=6°$, $v=0,015mm/s$; the effective deformation ε_f equals 15 and 130. The Cu microstructure with the increasing torsional frequency was presented in Fig.6a.

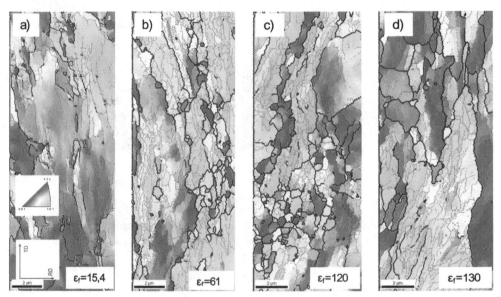

Fig. 6. EBSD maps of Cu after COT processing: a) f=0,2 Hz; b) f=0,8 Hz; c) f=1,6 Hz; d) 1,8 Hz at constant parameters: α =6°, v=0,015 mm/s i Δh=7 mm

The deformed copper with low torsional frequencies - 0,2 Hz is characterized mainly by the boundaries with a small misorientation angle. The boundaries like HABs are seen in the fragmentary outline.The Cu deformation with the increasing torsional frequency to 1,6 Hz, allows for the formation of the equiaxed structures having a great share of wide-angle boundaries (Fig.6b,c.). The maps obtained using the EBSD techniques show that the structures got during the deformation process where the torsional frequencies were – 0,8Hz and 1,6 Hz, are characterized by the considerable grain refining, especially when it comes to the deformation when the torsional frequency was 1,6 Hz. It was observed that the numerous grains had the size no bigger than 1 µm. The neighboring grains are characterized by comparable but yet diversified crystallographic orientation. The great part of the analyzed surfaces are the banding structures isolated by high-angle boundaries and elongated in accordance with the direction of the compression. The example that are shown here prove that even if the values of effective deformations ε_f=130 are high, the microstructure Cu is characterized by heterogeneity.

The deformation of the material when the torsional frequency is f= 1,8 Hz produces an inconvenient phenomenon of developing large grains which often exceed 1 μm.

This is proven by the results of the quantity tests of the grains and subgrains size (Fig.7), the area fraction of the grains with the misorientation above 15° and the size not exceeding 1 μm, (Fig.8) and misorientation angle (Fig.9) that are fulfilling the ultrafine-grained material criteria.

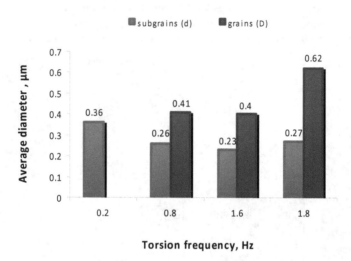

Fig. 7. The variation of the subgrain/grain size in dependence of changes in torsion frequency at constans: α =6°, v = 0,015 mm/s, Δh=7 mm

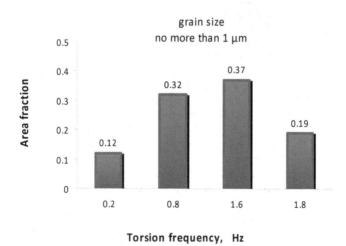

Fig. 8. The variation in the area fraction of ultrafine grain in dependence of changes in torsion frequency at constans: α =6°, v = 0,015 mm/s, Δh=7 mm

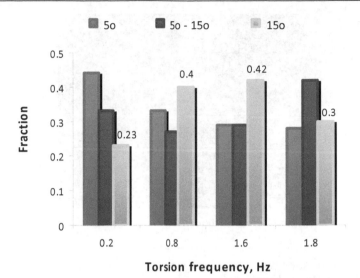

Fig. 9. The variation in the misorientation angle in dependence of changes in torsion frequency at constans: α =6°, v = 0,015 mm/s, Δh=7 mm

When the torsional frequency is 0,8 Hz and 1,6 Hz, the average diameters of the Cu grains have about 400 nm. A distinct increase of the average diameter of the grain is seen when the torsional frequency is increased from 1,6 Hz to 1,8 Hz. The grains have a size of about 600 nm. Roughly, it can be assumed that the subgrains are about 2 times smaller than grains. The area fraction of the grains to the magnitude of 1 μm, also appears to be beneficial for the torsional frequencies 0,8 Hz and 1,6 Hz (especially for 1,6 Hz) (Fig.8.). The Cu grains which have the ultrametric size occupy about 40% of the analyzed surfaces. The application of the high torsional frequencies during the deformation process allows for obtaining about 40% of the high- angle boundaries fractions (Fig.9). It means that more than a half of the deformed structure is dominated by the subgrains.

Fig. 10. Microstructure of Cu after COT deformation at: f=0,2 Hz, α =6°, v=0,015 mm/s and Δh=7 mm

Fig. 11. Microstructure of Cu after COT deformation at: 11. f=0,8 Hz, α =6°, v=0,015 mm/s and Δh= 7 mm

After the COT deformation with the torsional frequency f=0,2 Hz, the dislocation cell structure and the DDWs dislocation walls with a high density of dislocation are observed (Fig.10). The effects of arranging the dislocation structure are seen between the particular DDWs (Fig.10). In general, deformation carried out when the torsional frequencies are small makes the fraction of the narrow-angle boundaries (up to 5°) reaching almost 50% (Fig.9.) and the area fraction of the ultrafine-grains does not exceed 20% (Fig.8.).

It was observed that when the torsional frequency increased, the grain division into smaller volumes happened more often (Fig.11). The result of this is the generation of the greater amount of the LABs dislocation boundaries. A great amount of the tested areas of thin foils, shows that after the deformation with frequencies f= 0,8 Hz and f=1,6 Hz in particular, the dislocation polygonal walls and the grains with the HAB boundaries are created in the structure (Fig.12). The vacancy clusters visible in Cu, are the effects of the dislocation annihilation (Fig.13). The results of measuring the dislocation density (Fig.14) are also the proof of the above mentioned.

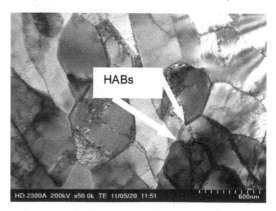

Fig. 12. Microstructure of Cu after COT deformation at: f=1,6 Hz, α =6°, v=0,015 mm/s and Δh= 7 mm

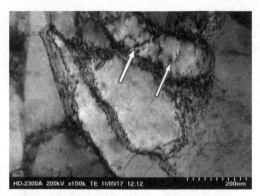

Fig. 13. Microstructure of Cu after COT deformation at: f=1,6 Hz, α =6°, v=0,015 mm/s and Δh=7 mm. The vacancy clusters are arrows marked

Fig. 14. Microstructure of Cu after COT deformation at: f=1,8 Hz, α =6°, v=0,015 mm/s and Δh=7 mm

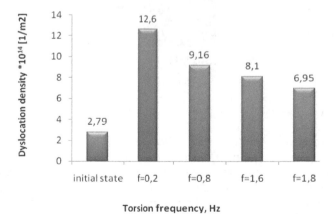

Fig. 15. The variation of the dislocation density in COTes samples with deferent value of torsion frequency and constant: α =6°, v=0,015 mm/s, Δh=7 mm

Together with the increase of the torsional frequency, the decrease of dislocation density is observed (Fig.15). When the frequency is 1,8 Hz the recovery begins to dominate the structure what has a negative influence on the material that is being refined (Fig.14). The transformation of dislocation tangles into polygonal boundaries and the dislocation annihilation, have no impact on raising the amount of the wide-angle boundaries fractions as well as on increasing the area fraction of the ultrafine-grains.

4.2 The compression velocity

In the COT process the compression velocity is the parameter which when increased, has an influence on highering the value of the compression force and on lowering the values of equivalent deformations. For example, the compression velocity 0,015 mm/s and 0,04 mm/s while the rest of the parameters is constant: 1,6 Hz, $\Delta h=7$, $\alpha=6°$; the effective deformation ε_f is 120 and 45. Applying even higher compression velocities i.e. v= 0,1 mm/s and v= 0,6 mm/s causes a significant decrease of effective deformation to the corresponding values $\varepsilon_f=12$ and $\varepsilon_f=2$.

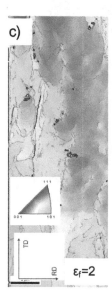

Fig. 16. EBSD maps of Cu after COT processing: a) v=0,04 mm/s; b) v=0,1 mm/s; c) v=0,6 mm/s; at constant parameters: α =6°, v=0,015 mm/s i $\Delta h=7$ mm

The deformation realized when the compression velocity is increasing from 0,015 mm/s to 0,04 mm/s while the rest of the parameters is constant: 1,6 Hz, $\Delta h=7$, $\alpha=6°$; has an positive influence on the grain refining as well as on the increase of the homogeneity of the structure (compare Fig.6c and Fig.16a). The structures in the prevailing part of the areas are equiaxed, the grains bigger than 1 μm are not produced, the amount of high-angle boundaries is high. Whereas, the further increase of this parameter delays the reduction of the grain size and the creation of high-angle boundaries (Fig.16 b,c). During the deformation with the velocity of 0,1mm/s the effects of the structure deformation are clearly visible (Fig.16b), but when the compression has the velocity of 0,6 mm/s only the fragments of the near-angle boundaries attest to the insignificant deformation of the material (Fig.16c). A significant reduction of the deformation realized by the increase of the compression velocity, does not lead to the grain refining.

When it comes to the Cu refined-grain, the juxtaposition of the structural changes for two velocities: 0,015 mm/s and 0,04 mm/s as well as of different torsional frequencies is quite interesting (Fig.17). That is why the further structural analysis involved comparing some selected ways of deformation.

Comparing the EBSD images obtained for Cu after the deformation with the compression velocity of v= 0,015 mm/s and v= 0,04 mm/s, the constant torsional frequency 1,6 Hz and the rest of the parameters also being constant (Fig.6c and Fig.16a), it is seen that the deformation happening at higher compression velocity is an effective way of grain-refining. It is also proven by the results of the quantity tests shown in the Figs.18-20. The average diameter of the Cu grain after the increase of the compression velocity, decreases from 400 nm to 300 nm (Fig.18). The average diameter of the subgrains also decreases from 230 nm to 180 nm (Fig.18). The area fraction of the grains up to 1 μm constitute 60% which is advantageous, when the torsional frequency is 1,6Hz and the compression velocity is v= 0,04 mm/s (Fig.19).

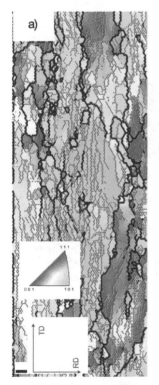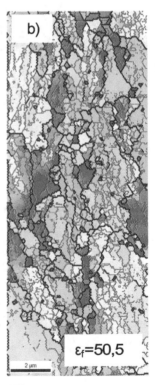

Fig. 17. EBSD maps of Cu after COT processing: a) f=0,8Hz; b) f=1,8Hz; at constant parameters α =6°, v=0,04 mm/s and Δh=7 mm

When the compression velocity is higher, the high-angle boundaries are easily created and they constitute about 50% of the analyzed Cu areas (Fig.20).

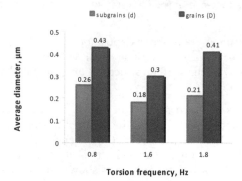

Fig. 18. The variation of the subgrain/grain size in dependence of changes in torsion frequency at constans α =6°, v = 0,04 mm/s, Δh=7 mm

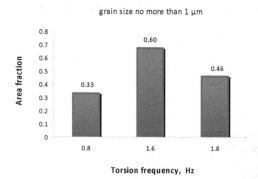

Fig. 19. The variation in the area fraction of ultrafine grain in dependence of changes in torsion frequency at constans: α =6°, v = 0,015 mm/s, Δh=7 mm

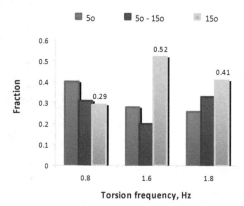

Fig. 20. The variation in the misorientation angle in dependence of changes in torsion frequency at constans: α =6°, v = 0,015 mm/s, Δh=7 mm

Comparing the structure of the deformed samples when the torsional frequency is lower – 0,8 Hz the compression velocities are as mentioned, it is visible that the Cu samples deformed with the higher compression velocity maintain greater fraction of the near-angle boundaries than the samples deformed with the lower compression velocity. The deformation realized when the compression velocity is v= 0,015 mm/s and v= 0,04 mm/s does not cause any significant changes when it comes to the size of the grains and subgrains (Fig.7 and Fig.18). The differences in the area fraction of the grains up to 1 μm were not observed (Fig.8 and Fig.19).

In the case of the deformation when the compression velocity was 0,04 mm/s and the torsional frequency was 1,8 Hz, it was observed that Cu structure was significantly smaller than after the deformation with the compression velocity equaling v= 0,015 mm/s (Fig.6d and Fig.17b). The area fraction of the grains up to 1 μm are also greater (Fig. 19) as well as the high-angle boundaries fractions (Fig.20).

STEM micrographs (Fig.21) evidently demonstrate that deformation at higher value of v parameter leads to generating banded structure with low angle grain boundaries and high value of dislocation denity (Fig.22).

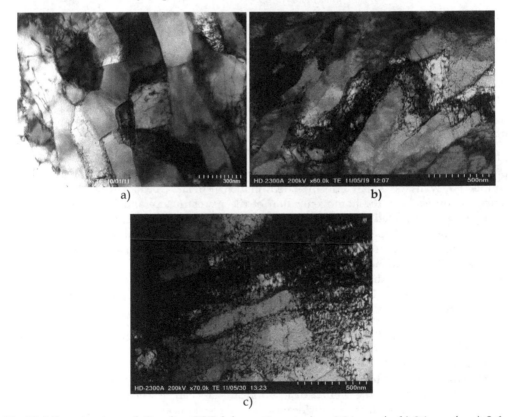

Fig. 21. Microstructure of Cu after COT deformation at: a) v=0,04 mm/s, b) 0,1mm/s, c) 0,6 mm/s and constans: f=1,6 Hz, α =6° and Δh=7 mm

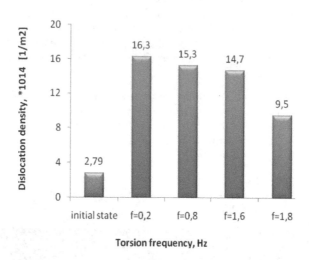

Fig. 22. The variation of the dislocation density In COTes samples with deferent value of torsion frequency and constant α =6°, v=0,04 mm/s and Δh=7 mm

4.3 The absolute deformation

The absolute deformation is the parameter, the increase of which has an impact on the increase of the value of ε_f parameter. Acording with literature, the heterogeneity is connected with too small effective deformation [Kuziak, 2005]. Using larger effective deformation leads to the homogenization of the microstructure and larger grain refining. Analyzing the selected samples of the size distribution of the subgrains (Fig.23), grains (Fig.24), and the shape indicator (Fig.25) using the EBSD technique, it is seen that the microstructure after applying a larger absolute deformation, is characterized by greater homogeneity. The homogenous character of the microstructure is indicated by the narrow spectrum of the size distribution in subgrains/grains and shape indicator. In the samples which are deformed to lower values of the absolute deformation, the great diversity of grain and subgrain size is observed.

The effect connected with the accumulation of the deformation through the increase of Δh is, above all, the grains fragmentation by the generation of the dislocation boundaries. The structures obtained after the absolute strain Δh =3mm, are characterized by the initial stage of creating the DDWs boundaries (Fig.26). In the inside of the areas that are divided by the dislocation boundaries, the cellular dislocation structure can be seen (Fig.26). After applying the deformation Δh =7 mm, the dislocation boundaries indicate a clear contrast proving the accumulating within the dislocation boundaries. A frequent intersection of the dislocation boundaries is observed (Fig.27).

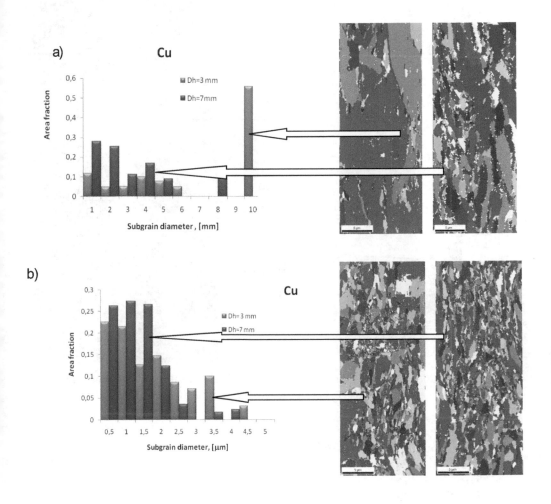

Fig. 23. Area fraction subgrains distributions of COTed Cu samples at: Δh= 3mm, Δh= 7mm; and at constant parameters: a) f=0,8 Hz, α =6°, v=0,04 mm/s , b) f=1,6 Hz, α =6°, v=0,04 mm/s

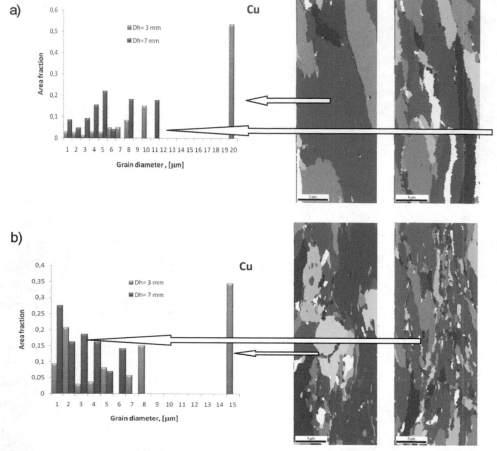

Fig. 24. Area fraction subgrains distributions of COTed Cu samples at: Δh= 3mm, Δh= 7mm; and at constant parameters: a) f=0,8 Hz, α =6°, v=0,04 mm/s , b) f=1,6 Hz, α =6°, v=0,04 mm/s

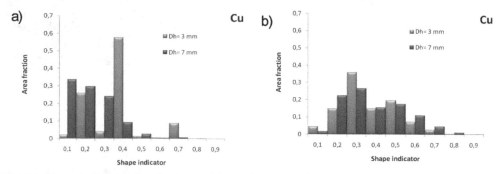

Fig. 25. Shape indicator distributions of COTed Cu samples at: Δh= 3mm, Δh= 7mm; and at constant parameters: a) f=0,8 Hz, α =6°, v=0,04 mm/s , b) f=1,6 Hz, α =6°, v=0,04 mm/s

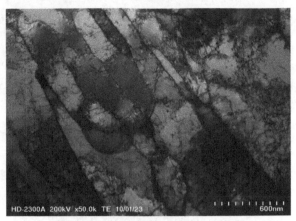

Fig. 26. Microstructure of Cu after COT deformation at: f=1,6 Hz, α =6°, v=0,04 mm/s and Δh=7 mm

Fig. 27. Microstructure of Cu after COT deformation at: f=1,6 Hz, α =6°, v=0,04 mm/s and Δh=7 mm

The above data shows that for the effective structure refining, it would be beneficial to lead the deformations with:

- the torsional frequency 0,8 Hz and 1,6 Hz. The large equivalent deformations realized by the increase of the torsional frequency ranging from 0,8 Hz to 1,6 Hz, have an impact on the increase of the misorientation value between the created grains (Fig.20), on the increase of the area fraction of the grains with the average diameter up to 1μm (Fig.19), on the creation of the equiaxed grains and subgrains in a range of the ultrametric sizes (Fig.18).
- the high value of the absolute deformation (Δh=7 mm). The increase of the effective deformation by means of the increase of the absolute strain has an influence on the raise of the high-angle boundaries fraction, the increase of the defects, and the generation of the dislocation boundaries which are subject to the mutual intersection (Fig.26,27). The microstructures are characterized by great homogeneity (Figs.23-25).

- the low compression velocity up to approximately 0,04 mm/s. The deformation when the compression velocities are higher than 0,04 mm/s has no positive influence on refining a grain to the ultrametric level (Fig.16b.c). However, the deformation proceeding when the compression velocities are very slow, fosters the structure recovery processes (Fig.6d).

Taking into account the most convenient conditions of the Cu grain refining process it should be stated that:

- the average diameter of the Cu grains and subgrains equals correspondingly about 300 nm and 200 nm.
- the area fraction of the grain which has the ultrafine-grained size, is 60%
- the fraction of high -angle boundaries (HAB) reaches up to about 50%

Despite the majority of the literature data does not discuss the created grain/subgrain separately but rather give the overall values, it should be stated that the obtained are comparable with the literature data [Dobatkin, 2007; Dalla Torre, 2004].

Less attention is paid to determine the fraction of high angle boundaries (HAB) while describing the structural effects. Only the work of Richert [Richert, 2006] and Dobatkin [Dobatkin, 2007] proved that the fraction of high angle boundaries (HABs) in Cu is above 50%. It is difficult to relate the obtained data concerning the area fraction of the ultrametric grains to the literature data because the available literature does not give such results.

The demonstrated decrease of a dislocation density with the increasing deformation, argues for the recovery processes accompanying deformation. The literature also tells about the decrease of the dislocation density after obtaining large deformations [Dalla Torre, 2004].

5. Copper mechanical properties after compression with oscillatory torsion

The measurements of hardness for Cu which was subjected to compression with oscillatory torsion, shows that the changes of hardness depends both on the compression velocity and torsional frequency (Fig.28). The growth of the torsional frequency causes the gradual decrease of hardness. However, the increase of the compression velocity causes the increase of hardness. When the torsional frequency was f= 0,2 Hz, the levels of hardness were the highest - even 130 HV0,2. The material deformed when the values of f parameter are low, does not create the grain/subgrain structure but is only characterized by numerous dislocation tangles. Thus, a large dislocation density is responsible for high hardness. The growth of the effective deformations ε_f caused by the increase of the torsional frequency does not contribute to the increase of hardness. A slow decrease of hardness that was observed is connected with the reduction of the dislocation density resulting from the recovering processes that it undergo. The data shown in the Fig.29,30 suggests that the refined grain of Cu leads to 1,5 fold increase of the ultimate tensile strength (UTS) when compared with the initial stage. It was also stated that the highest increase of the mechanical properties accompanies the deformed material when the compression velocities are higher. In general, the increase of deformation realized through the increase of the torsional frequency does not increase the UTS and yeld stress (YS) of the tested materials.

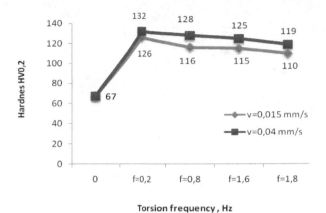

Fig. 28. Hardnes ploted as a function of torsion frequency and compression velocity; constants rest parameters: α =6°, Δh=7 mm

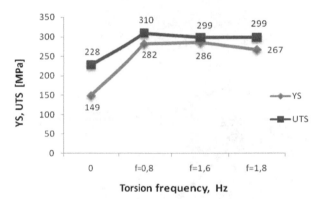

Fig. 29. YS and UTS ploted as a function of torsion frequency; constants rest parameters: v=0,015 mm/s, α =6°, Δh=7 mm

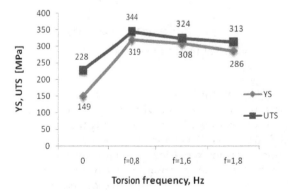

Fig. 30. YS and UTS ploted as a function of torsion frequency; constants rest parameters: v=0,04 mm/s, α =6°, Δh=7 mm

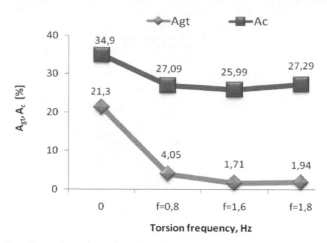

Fig. 31. A_{gt}, A_c ploted as a function of torsion frequency; constants rest parameters: v=0,015 mm/s, α =6°, Δh= 7

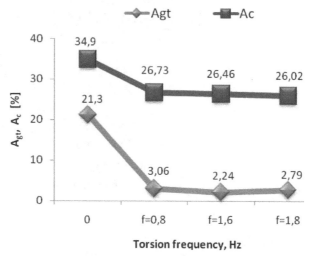

Fig. 32. A_{gt}, A_c ploted as a function of torsion frequency; constants rest parameters: v=0,04 mm/s, α =6°, Δh= 7

The reason of low ductility of the material deformed by the SPD techniques is the localization of deformation. In the case of ultrafine-grained materials with the average diameter of the grain not exceeding 1000 nm, the elongations are usually bigger than in the case of nanograined materials. It is caused by the greater abilities of the material to cumulate the dislocations in grains what leads to higher velocity of the work hardening. The Cu plasticity expressed by the elongation to the rupture A_c is lower in comparison to the initial state and does not change with the variable of deformation parameters (Fig. 31 and Fig. 32). However, the total uniform elongation A_{gt} reaches value of about 2% (Fig. 31 and Fig. 32).

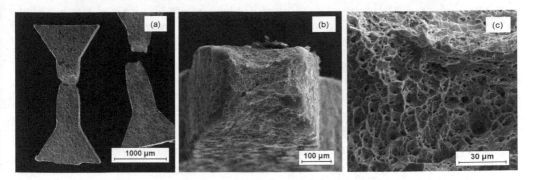

Fig. 33. Example fracture of the specimens 4 subjected to tension, b) necking near fracture section, c) fracture surface

Cracking of the materials after the tensile test happens in the areas of the measured parts (Fig. 33a). The material loses its stability as a result of the deformation localization in form of a neck which concentrates almost whole plastic deformation. In connection with that, the uniform elongation is negligible, however, the (YS) is close to the (UTS). The fracture surface prove the ductile character (Fig. 33b,c).

The mechanical properties of Cu after the deformation ensures the best grain size reduction is when f=1,6 Hz, Δh=7, α=6°, v=0,04 reaches values of about: YS =308 MPa and UTS = 324 MPa. The average UTS of Cu after the ECAP deformation is about 400 MPa [Kurzydłowski, 2004], and sometimes even higher values were noted [Beresterci, 2008]. The plastic properties, especially the uniform elongation for Cu, reaches the values of about 2%. This is the value comparable with the literature data [Kurzydłowski, 2004].

6. The mechanism of producing ultrafine-grained structures after the COT deformation

A lot of place in the structural tests is devoted to the mechanisms of grain-refining with the help of large SPD deformations. This matter is interesting particularly because of the application of different materials and different SPD techniques. The most important conclusions taken from the researches on the mechanisms of grain-refining are as follows:

- The fragmentation of grain takes place when the dislocation boundaries taking different forms are generated.
- The deformation is accompanied by the processes of dynamic recovery or even recrystallization [Kaibyshev, 2005, Wang, 2003]. An example of a material in which the grain-refining is the effect of deformation and of the „extended recovery" is aluminum. Whereas, in copper recrystalized grains are visible in the background of the deformed structure.
- The increase of misorientation happening thanks to the rotation of the grains boundaries and is a diffusion- controlled process. It is based on annihilation and absorption of the dislocation through the grains boundaries.

Hughnes and Hansen [Hughnes, 2000] presented the concept of microstructure evolution for the classic techniques of deformation which is based on the generation of dislocation boundaries which lead to the division of the initial grains into smaller volumes.

The proposed conception of the structure evolution is also characteristic for SPD techniques because many researchers dealing with refining the grain using large plastic deformations, introduce to the description of the structure the terminology in a form of the shear bands or dislocation layers. In general, there are three main mechanisms of the material structure-refining that are known:

- the production of new grains takes place thanks to a gradual increase in misorientation of dislocation boundaries as a result of absorption of new dislocations created during the deformation [Xu, 2005],
- the fragmentation of grains takes place thanks to the generation of the shear bands [Richert, 2006],
- the fragmentation of grains takes place thanks to the production of new grains as a result of the continuous recovery or continuous recrystallization [Kaibyshev, 2005, Wang, 2003] .

The refining of the copper grain after the COT deformation, happens as a result of the generation of dislocation boundaries, which together with the growth of deformation transform themselves into ultrafine-grained structure. The introductory stage of the grain-refining is the creation of DDWs dislocation walls the misorientation of which reaches even a few degrees and they stretch along the considerable fraction of a grain, separating blocks of dislocation cells (CBs) (Fig.34). Within the boundaries a high density of dislocation is accumulated (Fig.34).

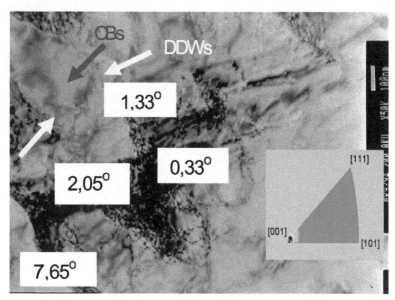

Fig. 34. Microstructure of Cu after COT deformation at: f=0,2 Hz, v=0,015 mm/s, α=6°, Δh=7 mm

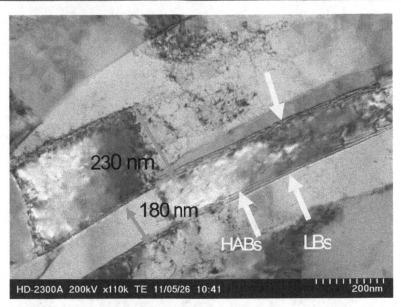

Fig. 35. Microstructure of Cu after COT deformation at: f=0,8 Hz f=0,2 Hz, v=0,015 mm/s, α=6°, Δh=7 mm

The growth of deformation causes the transformation of the DDWs dislocation walls into lamellar boundaries (LBs) which has larger misorientation (sometimes above 15° - HABs) which resembles long subgrains (Fig.35) Inside of the lamellar boundaries, the dislocation structure is regular and singular dislocation cells are usually noticeable. Moreover, the distance between the lamellar areas decreases (Fig.35). It can be assumed that the accumulation of deformation induces not only the creation of new dislocation boundaries and high angle boundaries but above all, it induces the crossing of dislocation boundaries. This phenomenon of intensive boundaries crossing (Fig.36) is a result of activating the subsequent slip systems while the process of deformation. The places where the dislocation boundaries are crossing induce the generation of almost equiaxed subgrains/grains (Fig.37). As a result, the large deformation increases the share of grains/subgrains boundaries and misorientation scattering (Fig.6) This, in turn, leads to the formation of the grains with HABs boundaries (Fig.37). The result of EBSD test shown in the Fig.6 and Fig.17 also constitute the confirmation of the tests. On the basis of EBSD tests it was proved that in a great deal of cases elongated neighboring grains remain in crystallographic compatibility. In the case of fine, equiaxed grains, the orientation was accidental (Fig.6).

The high-angle boundaries (HABs) marked in the Fig.36, 37 have bulges characteristic for the dynamic recrystallization and they can indicate the migration of high-angle boundaries. In many works it was proven that the migration of the grain boundaries created as a result of SPD process as an effect of dynamic recrystallization [Kaibyshev, 2005; Wang, 2003]. The sequence of figures registered during the rotation of the sample at a given angle indicates that the bulges of HAB boundaries are not the effects of the boundary migration but of the mutual superimposing of the boundaries which are in one crystallographic orientation in a given microarea (Fig.38).

This means that the creation of ultrafine-grained structures using the COT method is not determined by the process of recrystallization.

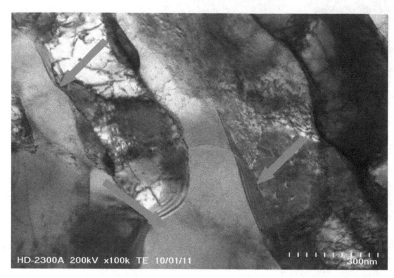

Fig. 36. Microstructure of Cu after COT deformation at: f=0,8Hz, v=0,015 mm/s, α=6°, Δh=7 mm

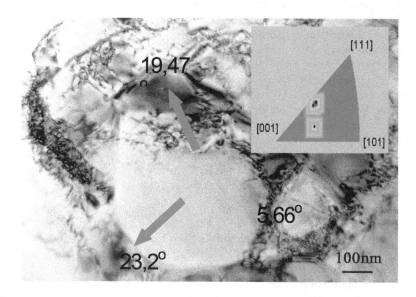

Fig. 37. Microstructure of Cu after COT deformation at: 1,6 Hz, v=0,015 mm/s, α=6°, Δh=7 mm

In order to trace the way in which the low-angle boundaries (LABs) change into HABs boundaries, a series of structural tests was carried out in which the Kikuch diffraction was used. Some examples are presented in the Fig. 39. Defining of the local orientations and crystallographic misorientations of particular areas, allowed to formulate the mechanism in which the high-angle boundaries are formed. The examples presented in Fig.39 suggest that the recovery of Cu caused by EBU happens relatively slow. That is why the dislocations generated during the deformation do not easily annihilation. The cellular type of the dislocation structure remains even if a large deformation is used. The EBSD tests show that the fraction of LABs boundaries is still present even after a large deformation (Fig.9,20).

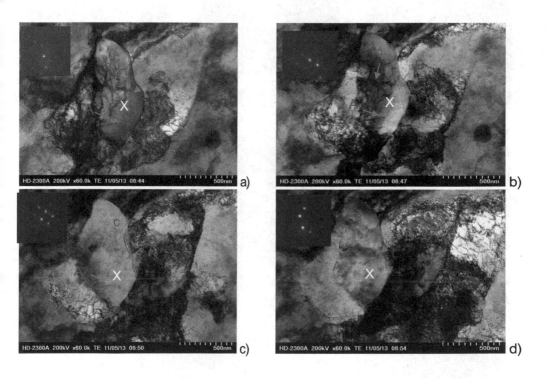

Fig. 38. Microstructure of Cu after COT deformation at: 0,8 Hz, v=0,15, α=6°, Δh=7 mm. ABC marked grains are wisible after changes in sample rotation : a) α= -3,9°, b) α= -6,2° , c) α= -9,6 °, d) α= -10,2° . Diffraction patterns taken from X

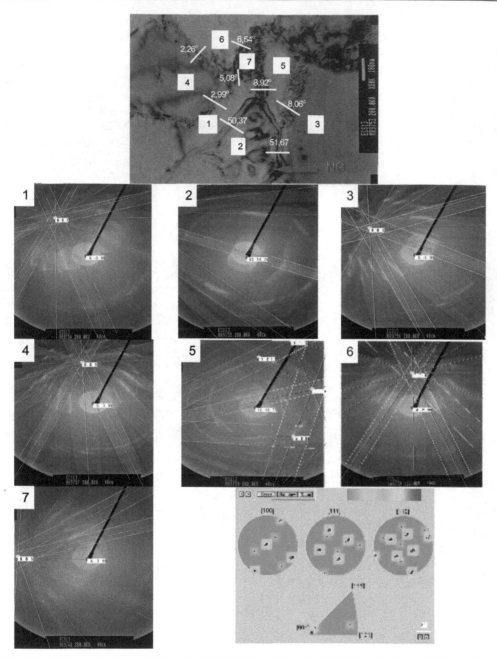

Fig. 39. a) Microstructure of Cu after COT deformation at: f= 1,6 Hz, α°=6, v = 0,04 mm/s. High misorientation of nonequilibrium grain boundary in grain marked 2.; b) Kikuchy difractions with solutions. Numbers 1-7 in Fig. 38a coresponds Kikuchy patterns 1-7 ; c) orientation of analized area

The clusters of dislocation are usually created near the boundaries what causes nonequilibrium state of the grains boundaries. A weak and heterogeneous diffractional contrast inside of the grains indicates a high level of the interior stress in the grain boundaries (Fig.39). Such boundaries are high-angular but almost invisible.

The structural analysis presented here show that the dominating mechanism of forming the high-angle grains in Cu is the growth of misorientation in dislocation boundaries. This, in turn, happens as an effect of absorbing dislocations to the grains boundaries during the deformation process. The nonequilibrium boundaries presented in the Fig.39 are formed from the absorption of numerous dislocations as a result of glide and climb [32].

It was proven that together with the growth of deformation, the fraction of the boundaries increases gradually from 5°-15° (up to approx. 30%) and the fraction of HABs boundaries raises above15° (to about 40%) (Fig.6). Approximately, on a constant level the fraction of narrow-angle boundaries LABs is formed and it does not exceed 30%. When the process parameters are changed (Fig.20) the increase of the HABs fraction to 50% is visible, but also the fraction of boundaries form 5°-15° was reduced to approximately 20%.

The LABs boundaries still constituted 30%. It can mean that the boundaries with misorientation from 5°-15° transform themselves into the grain boundaries. Because of the fact that the boundaries with an average misorientation (5°-15°) are the nonstable boundaries [Cao, 2008], as a result of the deformation they are transformed into high-angle boundaries.

7. Summary

The chapter concentrates on the possibility of a structure forming and on the properties of the metallic materials by means of using the method of compression with oscillatory torsion. The issues that were introduced have a great technical importance because it refers to the formation of ultrafine-grained structure by application of the unfamiliar SPD technique. Because the intensive research on generating new, more economical methods of producing ultrafine-grained materials, the presented matter coincides with the current state of research conducted in many scientific centers.

The aim of the tests in which the COT deformation method was used, was to obtain maximal refining of the grain below 1μm. The structural investigations showed that using the COT method, the grains of the Cu can be refined to ultrafine-grained. The intensity of grain-refining depends on the value of the effective deformation, however, the deformation path (the selection of the deformation parameters) is a decisive factor.

Using the different deformation parameters in process the presence of different phenomena that were controlling the growth of the microstructure. The increase of deformation (realized through the Hz growth) causes a progress in the grain refinement. The deformation conducted when the torsional frequencies were 0,8 Hz and 1,6 Hz is the most beneficial for obtaining the most refined grain. Using a significantly higher torsional frequency -18 Hz during the deformation caused considerable restrictions in the grain-refining because of the intensive recovery which begins to dominate over the deformation process.

It can be noticed that a relatively small change of the compression velocity, had an impact on considerably greater refining of the structure. The acceleration of the deformation process by increasing the compression velocity, causes the delay of the structure recovery process. What is seen through the changes of dislocation density for example. Despite reducing the effective deformation, the progress in the structure refining is observed what denies the results described in the literature [Dalla Tore, 2004; Wang, 2003]. However, the next increase of the compression velocity from 0,1 mm/s and 0,6 mm/s does not foster grain-refining. It should be explained by the fact that the effective deformation is too small.

The main effect of deformation is the increase of the structure's homogeneity. The homogeneity is obtained mainly by the increase of the height parameter Δh.

When the process parameters are as follows: the torsional frequency ranges from 0,8 Hz to 1,6 Hz; Δh=7mm, and the compression velocity ranges from 0,015 to 0,04 mm/s; the maximal refining of the copper grain is obtained but also:

-       the average diameter of the grain/subgrain correspondingly 200 nm and 300 nm.
-       the share of high-angle boundaries is about 50% and
-       the fraction of the ultrafine grains is about 60%. In spite of using large equivalent deformations, narrow-angle boundaries up to 15° constitute a significant fraction in both materials. The fraction of high-angle boundaries does not exceed 70%. This means that the thermal stability of the structures formed in such a way is sufficient [Raab, 2004; Lugo, 2008].

The result of mechanical tests were discussed in details for selected schemes of deformation in which the effects of deformation were visible the most. The influence of grain refining on the mechanical properties of the grain, was determined on the basis of the hardness measurements and the tensile test of the micro samples. The results of the mechanical tests that were carried out, indicate clearly that the changes taking place in the structure and the grain refining had a positive influence on the mechanical properties of the tested materials. However, the uniform elongation is not strongly dependent on the grain refining.

The difference in the properties that were observed, mainly result from using different schemes of deformation.

The results of structural investigations were the basis for determining the mechanisms of grain refining and forming high-angle boundaries. It was proven that the formation of ultrafine grains during the COT deformation is based on the general mechanisms of forming the dislocation boundaries. The process of grain refining proceeds by the generation of the LABs and HABs boundaries. In the introductory stage of deformation the dislocation boundaries are formed which are perpendicular to the direction of the compression force. The formation of the dislocation boundaries which proceed in such a way, suggests that in the initial stage of deformation it is mainly the compression that initiates the process of the grain refining. The non-directional process of deformation (the introduction of an additional torsion causing the change in the direction of loading) leads to the deformation of the material in more and more numerous systems of glides and to the new dislocations having influence on the active system of glides on the previously-generated dislocations. The effect of the introduced, additional loading is the increase in the number of the dislocation boundaries that cross mutually. When the effective deformation in the microstructure increases, the distances between the dislocation boundaries decrease – a new order of LBs dislocation boundaries is created which as a result of the thermal process activation (recovery), transform themselves into subgrains and grains. The process leads to the increase of misorientation of the grain's boundaries.

A significant role in forming the ultrafine-grained structure has the recovery process. Dislocations are rearranged, undergo annihilation and are also absorbed to the grain boundaries. Such a rebuilding of a dislocation structure causes the increase of the misorientation within the grain boundaries.

8. References

Pakieła, Z. (2009). *Mikrostrukturalne uwarunkowania właściwości mechanicznych nanokrystalicznych metali,* Oficyna Wydawnicza Politechniki Warszawskiej

Olejnik, L. et al. (2005). *Bulletin the polish academy of science, Technical sciences,* 53, pp. 413-423

Bochniak, W. et al. (2005). *Journal of Materials Processing Technology,* 169, pp. 44-53

Shaarbaf, M. et al. (2008). *Materials Science Engineering,* A473, pp. 28-33

Raab, G.J. et al. (2004). *Mater. Sci. Eng.,* A 382, pp. 30

Lugo, A. N. et al. (2008). *Materials Science Engineering,* A477, pp. 366-371

Kulczyk, M. et al. (2007). *Materials Science Poland,* 25, pp. 991-999

Pawlicki J.& F. Grosman. (2005). *Naprężenia uplastyczniające metali w warunkach złożonych obciążeń, Materiały XII Konferencji Informatyka w Technologii Metali,* Ustroń 16-19.01

Grosman F. & Pawlicki J. (2004). Processes with forced deformation path. New Forming Technology 2004. *Proceedings of the 1st ICNFT, Harbin Institute of Technology Press,* Harbin, China, Sep. 6–9

Rodak, K. & Goryczka T. (2007). *Solid State Phenom,* 130, pp. 111-113

Ungar, T. et al. (2001). *Materials Science Engineering,* A319-321, pp. 274-278

Kuziak, R.(2005). *Modelowanie zmian struktury i przemian fazowych zachodzących w procesach obróbki cieplno-plastycznej stali,* Instytut Metalurgii Żelaza, Gliwice

Dobatkin, S. V. et al. (2007). *Materials Science Engineering,* A462, pp. 132-138

Dalla Torre, F. (2004), *Acta Materialia,* 52, pp. 4819- 4832

Richert, M. (2006). *Inżynieria nanomateriałów i struktur ultra drobnoziarnistych,* Uczelniane Wydawnictwo Naukowo Techniczne, Kraków

Kurzydłowski, K. J. (2004). *Biuletin of the polish academy of sciences technical sciences,* 52, pp. 301-311

Beresterci, M. et al. (2008). *Metalurgija,* 47, pp. 295-299

Kaibyshev, R. et al. (2005). *Materials Science Engineering,* A398, pp. 341-351

Wang, G. et al. (2003). *Materials Science Engineering,* A346, pp. 83-90

Hughnes, D.A. et al. (2000). *Acta Materialia,* 48, pp. 2985-3004

Xu, Ch. et al. (2005). *Materials Science Engineering*, A398, pp. 66-76

Cao, W.Q. et al. (2008). *Materials Science Engineering*, A492, pp. 74-79

Permissions

The contributors of this book come from diverse backgrounds, making this book a truly international effort. This book will bring forth new frontiers with its revolutionizing research information and detailed analysis of the nascent developments around the world.

We would like to thank Prof. Krzysztof Sztwiertnia, for lending his expertise to make the book truly unique. He has played a crucial role in the development of this book. Without his invaluable contribution this book wouldn't have been possible. He has made vital efforts to compile up to date information on the varied aspects of this subject to make this book a valuable addition to the collection of many professionals and students.

This book was conceptualized with the vision of imparting up-to-date information and advanced data in this field. To ensure the same, a matchless editorial board was set up. Every individual on the board went through rigorous rounds of assessment to prove their worth. After which they invested a large part of their time researching and compiling the most relevant data for our readers. Conferences and sessions were held from time to time between the editorial board and the contributing authors to present the data in the most comprehensible form. The editorial team has worked tirelessly to provide valuable and valid information to help people across the globe.

Every chapter published in this book has been scrutinized by our experts. Their significance has been extensively debated. The topics covered herein carry significant findings which will fuel the growth of the discipline. They may even be implemented as practical applications or may be referred to as a beginning point for another development. Chapters in this book were first published by InTech; hereby published with permission under the Creative Commons Attribution License or equivalent.

The editorial board has been involved in producing this book since its inception. They have spent rigorous hours researching and exploring the diverse topics which have resulted in the successful publishing of this book. They have passed on their knowledge of decades through this book. To expedite this challenging task, the publisher supported the team at every step. A small team of assistant editors was also appointed to further simplify the editing procedure and attain best results for the readers.

Our editorial team has been hand-picked from every corner of the world. Their multi-ethnicity adds dynamic inputs to the discussions which result in innovative outcomes. These outcomes are then further discussed with the researchers and contributors who give their valuable feedback and opinion regarding the same. The feedback is then collaborated with the researches and they are edited in a comprehensive manner to aid the understanding of the subject.

Apart from the editorial board, the designing team has also invested a significant amount of their time in understanding the subject and creating the most relevant covers. They scrutinized every image to scout for the most suitable representation of the subject and create an appropriate cover for the book.

The publishing team has been involved in this book since its early stages. They were actively engaged in every process, be it collecting the data, connecting with the contributors or procuring relevant information. The team has been an ardent support to the editorial, designing and production team. Their endless efforts to recruit the best for this project, has resulted in the accomplishment of this book. They are a veteran in the field of academics and their pool of knowledge is as vast as their experience in printing. Their expertise and guidance has proved useful at every step. Their uncompromising quality standards have made this book an exceptional effort. Their encouragement from time to time has been an inspiration for everyone.

The publisher and the editorial board hope that this book will prove to be a valuable piece of knowledge for researchers, students, practitioners and scholars across the globe.

List of Contributors

Su-Hyeon Kim
Korea Institute of Materials Science, Republic of Korea

Dong Nyung Lee
Seoul National University, Republic of Korea

Yuriy Perlovich and Margarita Isaenkova
National Research Nuclear University "MEPhI", Russia

Vadim Glebovsky
Institute of Solid State Physics, the Russian Academy of Sciences, Russia

K. Sztwiertnia, M. Bieda and A. Kornewa
Polish Academy of Sciences, Institute of Metallurgy and Materials Science, Krakow, Poland

Kumkum Banerjee
Research and Development Department, Tata Steel Ltd., Jamshedpur, India

Toni T. Mattila and Jorma K. Kivilahti
Aalto University, Finland

Fritz Appel
Institute for Materials Research, Helmholtz-Zentrum Geesthacht, Geesthacht, Germany

R. Ebrahimi and E. Shafiei
Department of Materials Science and Engineering, School of Engineering, Shiraz University, Shiraz, Iran

Kinga Rodak
Silesian University of Technology Katowice, Poland

Printed in the USA
CPSIA information can be obtained
at www.ICGtesting.com
JSHW011439221024
72173JS00004B/862